THINKING ABOUT AMERICA'S DEFENSE

An Analytical Memoir

Glenn A. Kent

With
David Ochmanek
Michael Spirtas
Bruce R. Pirnie

Prepared for the United States Air Force

Approved for public release; distribution unlimited

 PROJECT AIR FORCE

This publication was sponsored by the United States Air Force under Contract FA7014-06-C-0001. Further information may be obtained from the Strategic Planning Division, Directorate of Plans, Hq USAF.

Library of Congress Cataloging-in-Publication Data

Kent, Glenn A., 1915–
    Thinking about America's defense : an analytical memoir / Glenn A. Kent ; with David Ochmanek, Michael Spirtas, Bruce R. Pirnie.
        p. cm.
    Includes bibliographical references.
    ISBN 978-0-8330-4452-5 (pbk. : alk. paper)
    1. Military planning—United States. 2. United States. Air Force—Officers—Biography. 3. Generals—United States—Biography. 4. Strategic forces—United States. 5. National security—United States. 6. United States—Defenses—Decision making. 7. United States—Military policy. I. Ochmanek, David A. II. Spirtas, Michael. III. Pirnie, Bruce, 1940– IV. Title.

U153.K466 2008
355'.033073—dc22

                                                                    2008025840

The RAND Corporation is a nonprofit research organization providing objective analysis and effective solutions that address the challenges facing the public and private sectors around the world. RAND's publications do not necessarily reflect the opinions of its research clients and sponsors.

RAND® is a registered trademark.

Cover design by Carol Earnest

© Copyright 2008 RAND Corporation

All rights reserved. No part of this book may be reproduced in any form by any electronic or mechanical means (including photocopying, recording, or information storage and retrieval) without permission in writing from RAND.

Published 2008 by the RAND Corporation
1776 Main Street, P.O. Box 2138, Santa Monica, CA 90407-2138
1200 South Hayes Street, Arlington, VA 22202-5050
4570 Fifth Avenue, Suite 600, Pittsburgh, PA 15213-2665
RAND URL: http://www.rand.org/
To order RAND documents or to obtain additional information, contact
Distribution Services: Telephone: (310) 451-7002;
Fax: (310) 451-6915; Email: order@rand.org

Preface

For those who have known and worked with Lieutenant General
(retired) Glenn Altran Kent over the years, it came as something of a
shock a few years ago when, in his late eighties, he (sort of) retired. That
is, he stopped coming into work every day. He has remained engaged
in U.S. national security affairs as an astute observer of policy, and
he continues to make important contributions to the work of his colleagues at the RAND Corporation and elsewhere. His colleagues still
turn to him for his insights on current work. But he is, alas, no longer a
daily presence. The void that this transition in General Kent's role created in the professional lives of his colleagues prompted us to approach
him about recording some of the high points of his career as a defense
analyst so that these invaluable lessons would not be lost. The result is
this volume.

This is not a memoir or a biography in the traditional sense of
these words. General Kent was not really interested in recounting the
events of his life, fascinating though they are. He was, however, willing and indeed eager to share what he has learned about analysis and
defense policymaking. Hence, he has produced what we call an analytical memoir, in which he shares his account of the most significant
issues with which he was involved over the course of his career—how
he saw each issue and its significance, how he conceptualized and
addressed the central analytical problems associated with the issue, and
how his work affected policy. Because General Kent's career in defense
began just before World War II and extended into the 21st century and
because he was intimately involved in many of the most salient national

security debates over the course of that span, to read this volume is, in many ways, to read an insider's history of key aspects of the Cold War and post–Cold War defense strategies of the United States.

Everyone who has worked with General Kent is indebted to him for the contributions he made to solving difficult, complex problems. Whether the task at hand was predicting the weather over Greenland in support of crews ferrying combat aircraft to England, setting the performance specifications for the Air Force's next frontline fighter aircraft, or outflanking the leadership of the Navy in support of the creation of the Single Integrated Operational Plan (SIOP) for U.S. strategic nuclear forces, General Kent always gave it his best. And his best was always very, very good. The stories collected in this volume are another tangible legacy of this uniquely creative, insightful, and influential man, and for this, we are again in his debt.

Project AIR FORCE

RAND Project AIR FORCE (PAF), a division of the RAND Corporation, is the Air Force's federally funded research and development center for studies and analyses. PAF provides the Air Force with independent analyses of policy alternatives affecting the deployment, employment, combat readiness, and support of current and future aerospace forces. Research is conducted in four programs: Force Modernization and Employment; Manpower, Personnel, and Training; Resource Management; and Strategy and Doctrine.

Additional information about PAF is available on our Web site: http://www.rand.org/paf/

Contents

CHAPTER FIVE

Modernizing Conventional Forces

CHAPTER SIX

Analytical Tools

Figures

Tables

Boxes

Acknowledgments

Glenn A. Kent

This book is a summation of the most significant work I have done over the course of my career. Along the way, I have had the good fortune to work with, learn from, and be helped by a host of highly capable people. Regrettably, it is not possible to give due credit to everyone to whom I owe it, but I am profoundly grateful to all of them. I do, however, wish to acknowledge several people who have played especially significant roles in my career.

First and foremost, my wife, Phyllis, my daughter, Kimberly, and my son, Cameron, merit special mention because of the extraordinary support and forbearance they have lent to me over many years. Anyone who has made a career in military or government service knows that such careers place special demands not only on those who serve but also on their families. My wife and children not only tolerated the stresses that my career placed on our family; they bore it all with grace and spirit, supporting me always.

I wish to give special mention to those who mentored me along the way. In roughly chronological order, they included Maj Gen Donald Yates, who was the Director of Research and Development on the Air Staff and my first boss in the Pentagon. General Yates taught me not to go with the crowd but to assess an issue with great care, get it right, and then have the courage of my convictions and make them known.

I had the good fortune to work for Lt Gen John Gerhart, who was the Deputy Chief of Staff for Plans and Programs on the Air Staff during the late 1950s and early 1960s. He was an extraordinarily effective leader who taught me just about everything there is to know about force planning. He combined exceptional vision with courage and bureaucratic savvy, making sure that the Air Force was always out in front in defining and fielding the capabilities that the nation would need. During the years that I worked for General Gerhart, I received invaluable assistance from (then) Lt Col Robert Lukeman (later, Maj Gen Lukeman), who worked in my division. Bob Lukeman was a man of many talents, but he made perhaps his greatest contributions during this period on issues relating to the management and evolution of the United States' stockpile of nuclear weapons. All of us in the Plans and Programs Directorate got lessons in doing quality staff work from (then) Col Richard Yudkin, who served as General Gerhart's executive officer. Dick was a staff officer par excellence who taught me how to lay out a policy issue concisely, present the facts of the problem, assess alternative courses of action, make a clear recommendation, and show how to counter the arguments of opponents. I handled countless such issues during my years there, including many that were formally considered by the Joint Chiefs of Staff, and I became rather adept at preparing, under short deadlines, briefings and papers that put forth the Air Force's positions clearly and in ways likely to gain the approval of high-level decisionmakers.

In 1961, I was privileged to spend a year as a fellow at Harvard University's Center for Internal Affairs. I made good use of this opportunity by, among other things, being sure to audit every course I could that was taught by Professor Thomas Schelling. Professor Schelling had done pioneering work in game theory and its application to the interactions between states engaged in conflict. His lectures taught me the importance of thinking about which side in a competitive or conflictual relationship has the "last move" at the operational and tactical levels— a concept that would serve me in good stead in my next assignment.

I was also extremely fortunate to have worked for Dr. Harold Brown when he served as Director of Defense Research and Engineering (DDR&E) under Secretary McNamara and later when he was Secretary of the Air Force. In those days, the "whiz kids" in Systems Analysis (later the Office of Program Analysis and Evaluation, or PA&E) got most of the press, but in my view, it was Dr. Brown who did the most to bring sound and sophisticated analyses to bear on the challenges of national security. From him, I learned that you should not back off from taking on large and complicated issues if they were important to national security.

I also forged a close and rewarding relationship with Dr. Brown's successor as Director of DDR&E, Dr. John Foster, who from time to time singled me out to run important projects when I was head of Development Plans at Air Force Systems Command. In every instance, thanks sometimes to doses of good luck, things worked out very well, much to the benefit of the Air Force and my own career. Another authority on force modernization for whom I have great respect and from whom I learned much was Dr. Alexander Flax, the Assistant Secretary of the Air Force for Research and Development from 1964 to 1969. Dr. Flax taught me a lot about physics in general and radars in particular.

I must pay special tribute to the two Chiefs of Staff for whom I worked directly as Chief of Air Force Studies and Analysis. Gen John McConnell first appointed me to the position in 1968, and Gen John Ryan kept me there through his tenure as chief (1969–1973). Both men placed uncommon trust in me and in the work of my people, affording us many opportunities to influence policies on a wide range of issues. In many ways, my time in Studies and Analysis working for these two chiefs was the high point of my professional career.

Whatever successes we enjoyed in those years were the result of having a great team of really smart people. Again, while it is not possible to name them all here, a few stand out as having played truly central roles. The first was Col Jasper Welch, whom I designated to be the Chief Military Analyst of Studies and Analysis. Jasper was an invaluable member of the team in Studies and Analysis. He had an uncanny ability to identify the key interactions inherent in complex problems, to quantify them, and to program computers to define these relationships. This last ability was quite rare in the 1960s and 1970s. Our computers were mainframes; the interface with them was with punch cards; and none of the software tools that we take for granted today existed.

I also relied very heavily on Lt Col Larry Welch (no relation to Jasper), who was my expert par excellence on matters relating to fighter aircraft and air operations. Larry had flown F-4s in Vietnam and had studied aeronautical engineering by going to night school at the University of Maryland. With this background, he put his brilliant mind to work on matters relating to fighter aircraft operations and modernization. Larry played a leading role in developing the TAC Avenger model, which allowed analysts to evaluate the relative performance of alternative aircraft designs in air-to-air combat. When the time came to define the performance parameters of the F-15, Larry was ready and played a dominant role in that effort. (Later, General Larry Welch went on to command the first operational F-15 wing at Langley Air Force Base and, in 1986, became Chief of Staff of the Air Force.)

Many of the issues on which Studies and Analysis did important work during my time there related to decisions about how to improve U.S. military capabilities for the defense of the Central Front in Europe. For these questions, we found that we needed a computer model that captured the main elements of conventional combat there (at least from the point of view of the airman), and so we created one. In this case, "we" means Maj Leon Goodson, Capt Scott Meyer, Capt Lou Finch, and Lt Ken Hinkel.[1] These four officers worked intensively on this problem and virtually nothing else for more than one year. When they were done, we had SABER GRAND—the perfect tool for helping gain insights on the important questions of the time. Armed with this tool, Studies and Analysis produced the most credible and influential work in the Department of Defense on airpower's contributions to the defense of the Central Front.

My more than 20 years at RAND have been extremely rewarding, and I would be remiss if I did not mention several colleagues with whom I have worked especially closely. First, I would like to thank Dr. Don Rice, former president of RAND, for inviting me to join the organization in 1982. I have also greatly valued my association with RAND's Natalie Crawford. Over the years, Natalie and I have collaborated on a range of projects, including the definition of the operational characteristics of the advanced tactical fighter (now the F-22). Throughout our long association, I have always appreciated her extraor-

---

[1]   Bob Lukeman, Jasper Welch, and Leon Goodson all went on to head Air Force Studies and Analysis after I left.

dinary knowledge of air operations and her love for the United States Air Force.

My time at RAND has yielded many other fruitful collaborations with highly capable colleagues. My coworkers on projects relating to first-strike stability and arms control merit special mention. Randall J. DeValk and, later, David E. Thaler did yeoman's work over several years assisting me in the development of analytical methods for quantifying first-strike stability and determining the effects of changes in U.S. and Soviet forces and postures on stability. David Thaler was also instrumental in helping me refine and articulate many aspects of the planning framework known as strategies to tasks. Edward L. (Ted) Warner made valuable contributions to our development of a new approach to arms control in the 1980s—one that would impose constraints on overall destructive capacity and not simply on numbers of launchers.

Finally, I wish to thank my colleagues who played important roles in helping to pull together this volume. Stephen T. Hosmer, a brilliant and exacting analyst in his own right, gets the credit (or blame) for convincing me to take on this project. He did so by conceiving it not as a memoir in the standard sense but rather as a collection of stories about some of the more-significant national security issues I worked on, and the analytical techniques I devised, to shed light on those issues. Besides getting the ball rolling, Steve was instrumental in helping to select the topics to be addressed, in organizing them into something resembling a coherent body, and in improving both the substance and the style of various iterations of drafts.

Special thanks also go to David Ochmanek, Michael Spirtas, and Bruce Pirnie, who provided research assistance, help in editing, and generally organizing and keeping track of all of the chapters that I wrote. David Ochmanek, a stalwart analyst and one of the more effective people at RAND in terms of getting things done and done right, was the fuel that made this engine run, for which I am ever grateful. Alex Hou did a careful review of the text, focusing on the mathematical formulations and the associated numbers. He also made creative suggestions about how to present more clearly several pieces of analysis described in this volume. Gens Larry Welch, Jasper Welch, and Leon Goodson each reviewed the entire draft, as did RAND colleague Timothy Bonds. Cynthia R. Cook shepherded the draft through the review process with care and diligence. Richard Hillestad, David E. Thaler, James T. Quinlivan, Russell D. Shaver, James Bonomo, and Joel Kvitky, all of RAND, also checked specific sections for accuracy and clarity. In spite of having many other responsibilities, Natalie Crawford graciously agreed to take responsibility for shepherding the draft through its final wickets. And Elizabeth Whitaker and Sarah Harting did their usual fine work of preparing and correcting the draft text. Steven Weprin coordinated final publication. Lauren Skrabala and Phyllis Gilmore expertly edited the text; Carol Earnest prepared the final art; and Erin-Elizabeth Johnson proofread the final copy. I am grateful to them all.

Glenn A. Kent
Lieutenant General, United States Air Force, Retired
Alexandria, Virginia
February 2008

Abbreviations

ABM	antiballistic missile
ACDA	U.S. Arms Control and Disarmament Agency
AFB	air force base
AFSA	Air Force Studies and Analysis
AFSC	Air Force Systems Command
AGZ	actual ground zero
AMPR	aeronautical manufacturer's planning report
AoA	analysis of alternatives
ATACMS	Army Tactical Missile System
AWACS	Airborne Warning and Control System
BMD	ballistic missile defense
BMEWS	Ballistic Missile Early Warning System
BMO	Ballistic Missile Office
CAG	conceivers' action group
CEP	circular error probable
CONEMP	concept of employment
CONEX	concept of execution
CP	correlation of power

CRAF	civil reserve air fleet
DARPA	Defense Advanced Research Projects Agency
DDR&E	U.S. Department of Defense Office of Research and Engineering
DE	damage expectancy
DGZ	designated ground zero
DIA	Defense Intelligence Agency
DoD	U.S. Department of Defense
DPM	draft presidential memorandum
DSP	Defense Support Program
erdel	a dummy term of measurement indicating xxx
FSB	fraction surviving of bombers; fraction surviving Blue (in certain equations)
FSR	fraction surviving Red (in certain equations)
GAO	U.S. General Accounting Office (now known as the U.S. Government Accountability Office)
GEO	geosynchronous orbit
GPS	Global Positioning System
HTK	hard-target kill
ICBM	intercontinental ballistic missile
IDA	Institute for Defense Analyses
IDAGAM	IDA Ground-Air Model
IOC	initial operational capability
IR	infrared
JCS	Joint Chiefs of Staff
JDAM	Joint Direct Attack Munition

JSTARS	Joint Surveillance Target Attack Radar System
JSTPS	Joint Strategic Targeting and Planning Staff
KP	kill potential
kt	kiloton
LEO	low earth orbit
LORAN	Long-Range Aids to Navigation
LWF	lightweight fighter
MIDAS	Missile Defense Alarm System
MIRV	multiple independently targetable reentry vehicle
MIT	Massachusetts Institute of Technology
MNS	mission needs statement
MVA	manufacturing value added
NATO	North Atlantic Treaty Organization
NAVSTAR	Navigation Satellite Timing and Ranging
nmi	nautical miles
OPLAN	operational plan
ORD	operational requirements document
OSD	Office of the Secretary of Defense
PA&E	Office of Program Analysis and Evaluation
PAF	Project AIR FORCE
PBV	post-boost vehicle
PI	probability of intercept
P_k	probability of kill
PPBS	Planning, Programming, and Budgeting System
PPI	polar projection indicator

R&D	research and development
RFP	request for proposal
RSIOP	a Russian equivalent to the U.S. SIOP
RV	reentry vehicle
SAB	Scientific Advisory Board
SAC	Strategic Air Command
SALT	Strategic Arms Limitation Talks
SAM	surface-to-air missile
SDI	Strategic Defense Initiative
SIOP	single integrated operational plan
SLBM	submarine-launched ballistic missile
SPO	system program office
SRAM	short-range attack missile
SSBN	ballistic missile submarines (formally, strategic submarine ballistic nuclear)
START	Strategic Arms Reduction Treaty
STRATCOM	U.S. Strategic Command
TAC	Tactical Air Command
TPP	Total Package Procurement
USAF	United States Air Force
USD(AT&L)	Under Secretary of Defense for Acquisition, Technology, and Logistics
USSR	Union of Soviet Socialist Republics
WSEG	Weapon System Evaluation Group

Creating Strategic Analysis

Thomas C. Schelling

In 1963, Colonel Glenn Kent of the United States Air Force, who was my guest at the Harvard Center for International Affairs, published an "occasional paper" of that center in which he looked at the question: If we were to have a limit of some kind on strategic missiles, what would be the most sensible limit? He argued that we should want both sides to be free to proliferate weapons in whatever dimension would reduce their own vulnerability without increasing the other side's vulnerability.

In those days, missile accuracies were poor, and megatonnage mattered more than today; big explosives, however, were less efficient than small ones because the lethal area was less than proportionate to the energy yield of the individual warhead. Kent proposed that the ideal magnitude to limit was the sum of the lethal areas covered by all the warheads in the inventory. This would be calculated as the sum of the two-thirds power of the yield of each weapon. In this formula, each party would be free to multiply smaller and smaller warheads on more and more missiles, thus becoming less and less vulnerable without acquiring any more preemptive attack capability.

He further calculated—and this is pure serendipity—that the weight-to-yield ratio went up as warheads got smaller; that the weight of the warhead would be roughly proportionate to the two-thirds power of the yield; and that, no matter how many warheads were on a given missile, the physical volume of the missile required to launch that weight would be approximately proportionate to that calculated index of lethality. And you could estimate the volume of the missile by looking at it from a distance! So monitoring would be easy and unobtrusive.

At the time, it was the neatest piece of strategic analysis I had ever encountered. Now, nearly half a century later, it still is. Later—and this must be some twenty-five years ago, when ballistic missile defenses were of renewed interest on account of President Reagan's "Strategic Defense Initiative"—Kent, by then retired from the Air Force and working at RAND, presented to a Harvard audience a dynamic analysis of the process of traversing an era of instability to arrive, one hoped, at a stable outcome. You will find that in an early chapter; I cannot condense it here. But it was an analysis I had never seen before, and I doubted anyone I knew could do it except General Kent.

He is still, at an age I won't try to guess, just as capable of articulate economical language and perceptive analysis as when I first met him at RAND. Instead of enjoying the comfortable retirement he deserves, he has provided us with a personal history of U.S. nuclear strategy from the same keen point of view that I had seen displayed long ago. When asked by his colleagues whether I'd consider doing a foreword to this personal review, I knew I owed it to Glenn Kent to express the pride I've always taken in having had some small influence in facilitating a unique career in a uniquely awesome profession.

Thomas C. Schelling
Lucius N. Littauer Professor of Political Economy, Emeritus,
Harvard University
Distinguished University Professor, Emeritus,
University of Maryland

Putting Analysis to Work

Harold Brown

I first met Glenn Kent, then a lieutenant colonel, in late 1954 or early 1955 at a meeting convened by Ramo-Wooldridge, the system designers for the Air Force ballistic missile program, to consider intercontinental ballistic missile designs and the integration of their reentry vehicles and nuclear warheads. (The meeting took place, perhaps appropriately, in a deconsecrated church on Aviation Boulevard near the Los Angeles airport.) Even in that brief session, Glenn's analytical skills were obvious and impressive. During the subsequent fifty-odd years, he has done more to illuminate the decision process on key Department of Defense (DoD) issues through analysis than any other individual. Concurrently, he has trained several professional generations of analytic thinkers in the Office of the Secretary of Defense (OSD), the Air Force, the Weapons System Evaluation Group (advising the Joint Chiefs of Staff and OSD), and the RAND Corporation. He has also induced a variety of senior military leaders and civilian officials (including me) to think more clearly and decide issues more rationally.

A model of analysis, and probably Kent's most influential one, was the study commissioned by Secretary of Defense Robert McNamara, that Glenn led in 1963–1964, when he worked in DoD's Office of Research and Engineering while I was Director of Defense and Engineering. It considered the prospects for limiting damage in an all-out nuclear exchange by examining the interaction between Soviet and U.S. strategic offensive nuclear forces and the effects of possible defensive forces and strategies on both sides. It made clear that the combination of the devastating effect of thermonuclear weapons, the vulnerability of urban-industrial society, and the ability of the offense to pick the nature of its tactics after the defense had deployed its elements, meant that neither the United States nor the Soviet Union could avoid national destruction in an all-out thermonuclear exchange. That result dictated that the basis for U.S. nuclear strategy through the end of the Cold War, despite the hopes held out for fallout shelters in the 1960s and the Strategic Defense Initiative in the 1980s, had to be the preservation of stable nuclear deterrence in the shadow of assured mutual destruction if deterrence failed. That in turn became the criterion for decisions on U.S. strategic force structure and for negotiations on strategic arms limitation and reduction, outlasting first-strike aspirations, the Strategic Defense Initiative, and finally the Soviet Union.

Kent's influence, however, goes beyond the effect of particular studies, through both his mentoring efforts and his style: simplicity and transparency of the model; clear choice of measure of merit as seen through the eyes of the decisionmaker; realistic inputs on costs and on technological issues; uncovering the key drivers of the outcome; and displaying the way in which the answers depend on the assumptions. This last allows, indeed encourages, the decisionmaker, whose choices the analysis is supposed to illuminate, to see how his or her policy preferences, instincts, and even prejudices affect the answers given by the analysis. Those policy elements remain as important, often decisive, factors in the decision, but the analysis, done in that way, allows them to affect the decision in an understood way, rather than by intuition alone.

The explosion of new programs in response to Sputnik and the Soviet ballistic missile program, and the collision of their projected costs with budget limits, had led Secretary of Defense Thomas Gates to look for ways to make choices during the closing years of the Eisenhower administration. But the value attached to analysis throughout DoD grew explosively in the 1960s because of McNamara's use in his decisions of the work of the Systems Analysis Office, led by Alain Enthoven. Alain and his immediate boss, Charles Hitch, the DoD Comptroller, had come to OSD from the RAND Corporation, where such analysis had been brought to a new level. In self-defense, the military services followed suit. Indeed, after the damage-limiting study demonstrated that competent analysis elsewhere could stand up to the work of the Systems Analysis Office, the Air Force, followed by the Navy and Army, set up their own units to carry out such analyses. The work of these offices both informed the decisions of their respective service secretaries and military chiefs and gave them useful material in their inevitable appeals of adverse decisions by the Secretary of Defense. How much decisionmaking improved overall is an open question, because there are many other factors at play aside from analysis: political factors, greatly increased and detailed congressional and public scrutiny, and employment issues, for example. In any event, by the late 1960s, such analysis of improved quality throughout, and generally accepted within, the DoD was an indispensable element in decisionmaking. Naturally, in 1968 Glenn Kent became the Assistant Chief of Staff of the Air Force for Studies and Analysis.

Equally or even more important, but less universally adopted, were Kent's attempts to introduce a new paradigm for innovation and modernization of forces. His "strategies-to-tasks" conceptual hierarchy properly insists on relating the combat systems embodied in innovation and modernization to the tasks whose execution is needed to achieve the operational goals of combatant commanders, and from there up through the hierarchy of military strategies, national security objectives, and U.S. interests. Instead, the use of the word *requirements* by operators uninformed by analysis and the pursuit of technology at or beyond the bounds of reality (which Kent has characterized by its dependence on the imaginary new element "unobtainium") continue to combine to encourage failed systems developments and horrendous costs. Indeed, if analysis is a signal success in DoD, the development and procurement system remains dysfunctional. Each effort at reform ends in the addition of a new layer of review and decision rather than a recasting of the system from the bottom up. It is a weak defense of the DoD system to note that it is a model of effectiveness by comparison with the rest of the federal government.

Analysis in support of decisionmaking began to spread beyond DoD to other cabinet departments late in the 1960s. Analogous activities were launched in the departments of Health, Education, and Welfare; Housing and Urban Development; and Transportation, for example, sometimes by alumni of the OSD Systems Analysis Office. The quality of analysis has often been lower, but that is not the principal difficulty faced by such efforts. The domain of a study to inform policy decisions on civil and domestic matters is likely to be multispectral in terms of players, advocates, and issues, and the issues are often more complex. There is not a single adversary, so a measure of merit is more difficult to agree on. It's not Red on Blue, Blue on Red; the Lanchester equations don't apply. Perhaps the toughest obstacle is that so many of the players deny that trade-offs among desiderata are appropriate; most advocates insist that, in these matters, *trade-off* is an obscene word. That makes optimal, or even acceptable, solutions hard to find.

Nevertheless, useful studies and analyses are being done in civil areas. At RAND, for example, where Glenn Kent has spent many years since his retirement from the military, policy-related analyses on health services, education, and energy have provided useful inputs to the public consideration of those issues and have occasionally even influenced decisions made in the Executive Branch, Congress, and state and local governments. The key element needed in those efforts, as in those on the national security side, is what Glenn Kent has brought to his studies and analyses: insight and integrity.

Harold Brown

Counselor, Center for Strategic and International Studies
Department of Defense Office of Research and Engineering, 1961–1965
Secretary of the Air Force, 1965–1969
Secretary of Defense, 1977–1981

Introduction

David Ochmanek, Bruce Pirnie, and Michael Spirtas

Glenn Altran Kent was a uniquely acute player in American defense policy in the second half of the 20th century. From 1957, when he joined the Weapons Plans Division of the Air Staff in the Pentagon, until his retirement from active duty in 1974, he was among the most perceptive and influential officers in the United States Air Force. For the next two decades, from his perch at the RAND Corporation, he published analyses on a broad range of topics that both shaped and raised the level of debates regarding the nation's security. A selected list of the issues in which General Kent played a decisive role is sufficient to give a sense of the scope of his influence:

- the inception of the single integrated operational plan (SIOP) governing the wartime employment of U.S. strategic nuclear forces
- the acknowledgement, in the early 1960s, of the dominance of strategic offensive nuclear forces and the subsequent abandonment by the mid-1960s of major efforts to field strategic defenses
- the conception of strategic nuclear arms control treaties as a means not only of constraining the destructive potential of the U.S. and Soviet nuclear arsenals but also of enhancing strategic stability by strengthening the survivability of those forces
- rigorous evaluations of the effects of deploying national missile defenses under the Strategic Defense Initiative (SDI)
- the development of military systems that have been central to the overwhelmingly successful U.S. military operations of the 1990s and beyond:
  - the F-15 fighter
  - the Airborne Warning and Control System (AWACS)

- the Joint Surveillance Target Attack Radar System (JSTARS)
- a variety of precision-guided munitions
- the Defense Support Program (DSP) satellite early warning system.

In addition to these issues, General Kent was personally involved in a host of other decisions that helped to shape the contours of U.S. national security strategy. Over the course of his career, he developed innumerable conceptual frameworks and analytical approaches that have lasting relevance to students and practitioners of national security policy. It is impossible to relate all these, but this volume is intended to tell the story of the most significant, providing sufficient information about the context and the content of his work so that future generations can adapt and apply the products of his creative genius to new problems that arise.

Glenn Kent was born on June 25, 1915, in Red Cloud, Nebraska. In 1918, his family moved to Manzanola, Colorado. Manzanola is a farming community on the Arkansas River in southeastern Colorado. This is hardscrabble territory even today, and in the drought years of the Great Depression, times were especially hard. Glenn grew up helping his father farm. In the mid-1920s, after several bouts of unfavorable weather, the family lost the farm and moved to a house in town.

Glenn was the valedictorian of his high school class (1932), which entitled him to a certificate that covered his tuition at any state college or state university. This, along with the earnings from a series of part-time jobs, allowed him to attend Western State College in Gunnison, Colorado, where he majored in mathematics. After graduating in 1936, Glenn taught math and chemistry at the high school in the small town of Hotchkiss in the mountains of western Colorado. He also coached the school's basketball team. In three years at Hotchkiss, Glenn turned the school's basketball team around from being a perennial loser to being a team with confidence and pride. He preached a style of play then known as "Romney basketball," which featured the fast break and a tenacious defense. Applying a philosophy that would serve him well later in his career, Glenn also taught his players to be aggressive, pointing out that "It's a foul only if the referee calls it."

In June 1941, Glenn joined the Army Air Corps as an aviation cadet. His decision to volunteer for the Air Corps was influenced by the reintroduction of the draft: He thought that he would prefer flying to serving in the infantry. However, because of an injury he had sustained to his ankle while playing basketball at Western State, he was declared ineligible for flight training. Instead, the Army Air Corps sent him to the California Institute of Technology (Caltech) in Pasadena, California, to study meteorology and fluid dynamics.

It was while Glenn was at Caltech that the Japanese fleet attacked Pearl Harbor. Within months, he was part of the vast buildup of the U.S. armed forces for World War II. In 1942, he was sent to the offices of Eastern Airlines in Atlanta to learn more about forecasting weather for flight operations.

After a short stay in Atlanta, he was posted to the weather detachment at Goose Bay, Labrador. He had been at Goose Bay for about a year when, by chance, he was joined at dinner one evening in the officer's club by a senior officer who was passing through Labrador on his way to Greenland. The officer's name was Bill Ford, known to many as "Wild Bill." Colonel Ford was an aviation pioneer who had made his reputation as a highly skilled captain for Trans World Airlines.

When wartime production began to crank up, the Air Corps started to ferry aircraft to England by the hundreds—B-17s, B-25s, B-26s, and others. Unfortunately, an underdeveloped infrastructure, a dearth of pilots trained to fly in poor weather, and vagaries in the weather in the North Atlantic conspired to make the journey across the Atlantic quite hazardous: Loss rates for the ferried aircraft at times approached 5 percent.

Determined to solve this problem, Gen Henry Harley "Hap" Arnold, commander of the Air Corps, offered Ford a direct commission as a colonel and the authority to have complete charge of running the ferry operation across the North Atlantic. Ford loved a challenge and accepted. On his way to Greenland to take command, he was on the lookout for promising officers to serve on his staff. He evidently saw something in Captain Kent that he liked, and before dinner was over, Ford offered Kent the job as the chief of the weather station at the base in Greenland known as BW-1. Glenn accepted (as he saw it, he had no choice) and was directed to go pack his bags and to be on the plane taking Colonel Ford to BW-1 that night.

Ford set about at once to improve every aspect of the system for ferrying aircraft to Europe. Among the most important steps he took were to expand the capacity of BW-1 and the base in Iceland to park and tie down aircraft that were in transit to England. He also went to great efforts to upgrade the communication systems and procedures governing the flow of aircraft through the system. And he set about to improve the reliability of weather forecasts by using weather observers in B-25 aircraft to report the flying conditions between BW-1 and Iceland just prior to the dispatch of aircraft in transit to Iceland.

Within a few months, the situation had improved dramatically: Loss rates plummeted to near zero even as throughput rose substantially. One reason for their success was that Ford had the authority to go directly to General Arnold when he needed something, and Arnold would make it happen. Another reason was hard work: Ford drove himself and his staff as if lives depended on their work—which, of course, they did. By observing Colonel (later General) Ford, Glenn saw what determined leadership and brains could do in the face of a complex problem.

After a brief postwar return to civilian life, Glenn was called back to what was soon to be the U.S. Air Force and was assigned to the Naval Postgraduate School, then located at Annapolis, Maryland. This gave him the opportunity to improve his quantitative skills. He characterizes the math curriculum at the school as "extremely rigorous." The washout rate was high, but Glenn made the cut.

At the end of the course, he was one of a dozen or so officers selected to go to the University of California at Berkeley to study radiological engineering, on the premise that the armed forces would have the responsibility for civil defense if the United States were attacked with nuclear weapons.

After his stint at Berkeley, Glenn was assigned to the Research and Development directorate of the Air Staff, then headed by the infamous Maj Gen Donald Yates. Yates was brilliant, irascible, eccentric, and tenacious. One of his management gambits was to announce to his staff a policy position that he privately disdained. Those who then spoke in favor of it in hopes of gaining favor with the boss were vilified and, sometimes, fired on the spot. "I say a lot of things," Yates would declare. "Your job is to tell me which ones are smart and which ones are stupid." Fortunately, Glenn avoided such entrapments and gained some respect from the general. In time, Yates began to rely on Major Kent to take the lead on a number of issues relating to nuclear weapons. This did not mean that Glenn was spared the occasional tirade.

One of the issues they grappled with and on which they ultimately prevailed was how to quash a plan, which many in DoD backed at the time, to develop a radiological area-denial weapon (see "Killing the Concept for an Area-Denial Weapon," pp. 123–126).

During this first tour of duty on the Air Staff, Glenn met Phyllis Horton, who had been a teacher of English in the high school at Richlands, Virginia. In 1953, they wed, and she would be his lifelong companion and inspiration.

After a few years on the Air Staff, General Yates arranged for Glenn to be assigned to the Air Force Special Weapons Center at Kirtland Air Force Base, outside Albuquerque, New Mexico. There, he was to be in charge of the Research Directorate. One of Glenn's responsibilities at Kirtland was to oversee the development of the nuclear-armed MB-1 rocket, later known as the "Genie." His stewardship of this program is chronicled in "Developing the MB-1 Rocket" (pp. 128–137).

From Kirtland, Colonel Kent went to Montgomery, Alabama, to attend the Air War College. While there, he volunteered for duty in the Air Staff's Directorate of Plans back in the Pentagon. He arrived back in Washington in 1956. Gen Thomas White was the Chief of Staff of the United States Air Force, and the Deputy Chief of Staff for Plans and Programs was Lt Gen John Gerhart. Glenn describes Gerhart as a visionary who never lost sight of the needs of the nation or of the Air Force's role in helping to meet them but who also knew how to get things done. Under him, Air Force Plans and Programs had its heyday. General White, who was charming and affable but not a "hands on" leader, trusted Gerhart implicitly. And because they had been associates in World War II, Gerhart was also on good terms with Gen Curtis LeMay, then commander of the powerful Strategic Air Command.

Glenn headed the division under General Gerhart that was responsible for developing the Air Force's positions on the type and number of nuclear weapons that the armed forces would order from the Atomic Energy Commission. As such, Glenn was deeply involved in the Joint Staff's deliberations on the fabrication and allocation of nuclear weapons. Because nuclear weapons were then at the heart of U.S. military strategy, decisions about their design, production, and allocation were critically important. They were also extremely contentious and evoked heated debates among the services. By marshalling rigorous and logical arguments, targeted analyses, and clever bureaucratic maneuvers, Glenn developed a reputation for prevailing on issues brought to the Joint Staff.

Soon, General Gerhart had Glenn involved in the full range of issues relating to the future of the Air Force. For example, long before the United States became heavily involved in Vietnam, General Gerhart recognized that the Air Force would need to improve its capabilities for conventional military operations. He saw that the threat of escalation to nuclear use would not be credible as a response to lower-level aggression and that so-called "small wars" would not be simply lesser included cases of "the big one." At a time when the Air Force's Tactical Air Command (TAC) was desperate to get into "the megaton business" as a way of ensuring its relevance, General Gerhart was developing plans for a TAC that would de-emphasize nuclear weapons and place top priority on acquiring fighter aircraft best suited to conventional conflict.

In 1960, while working for General Gerhart, Glenn scored one of his greatest coups. The Secretary of Defense at that time, Thomas Gates, was interested in finding a means for imposing more cohesion and integration on the nation's operational plans for employing nuclear weapons. Threatening the Soviet Union with a retaliatory nuclear attack had for some time been the centerpiece of U.S. efforts to deter a Soviet attack on the United States. Despite this, U.S. plans for nuclear attacks on the Soviet Union and the Warsaw Pact were not being developed in an integrated manner. Rather, they were developed by the individual regional combatant commands and were not well coordinated.

Charged by General Gerhart to involve himself in this problem, Glenn played a pivotal role in conceiving of a process by which a joint planning staff would develop an integrated plan. He called it the SIOP. Working with Gen Thomas Power, who was then commander of SAC, and Col George Brown, who was military assistant to Secretary Gates, Glenn helped to choreograph the process for gaining the secretary's approval of this concept (see "The Advent of the SIOP," pp. 22–30). One result of the SIOP triumph was that General Power, who had a reputation as fearsome as LeMay's, saw to it that Colonel Kent was eventually promoted to brigadier general, in 1963.

In 1961, Glenn was assigned to Harvard University, where he spent a year as a fellow at the Center for International Affairs. While there, he immersed himself in courses on strategy, economics, and game theory as taught by Dr. Thomas Schelling, later a Nobel laureate. Glenn also began work on a subject that would remain a focus for the rest of the Cold War: arms control and first-strike stability. Many of the seminars at Harvard that Glenn attended featured heated discussions. This proved to be good preparation for controversies that were yet to come (see Chapter Two).

In the summer of 1962, Glenn was assigned back to the Pentagon. This time, he was working not for the Air Force but in the Office of the Secretary of Defense (OSD). Specifically, he worked for Dr. Harold Brown, who was then the Director for Defense Research and Engineering (DDR&E). This was the heyday of Robert S. McNamara and the "Whiz Kids"—analysts brought to OSD from places like RAND. They were recruited in an effort to bring greater rationality to defense decisionmaking through the application of systems analysis. In Dr. Brown, Kent had a boss with a superb grasp of mathematics and defense planning. Dr. Brown also enjoyed the trust and confidence of Secretary McNamara. The Office of Systems Analysis was headed by Charles Hitch and was home to many of the whiz kids. A competition of sorts emerged between DDR&E and Systems Analysis.

Glenn had impressed Brown with some early work he did on issues relating to the defense of the United States against nuclear attacks. In time, Brown directed Kent—then a brigadier general—to analyze the utility of the full range of programs for limiting damage to the United States from a nuclear attack—offensive weapons and delivery systems, active defenses (air and missile), warning systems, and passive defenses (such as fallout shelters). This effort consumed General Kent's time for the better part of a year. The study's primary conclusion was that the U.S.–Soviet strategic balance was dominated by offensive systems and that any investment the United States might make in systems designed to limit damage from a Soviet attack could be overwhelmed at less expense by a larger attack. This insight led to a shift in policy that emphasized ensuring a secure second-strike capability over limiting damage and laid the conceptual foundation for two-sided arms control (see Chapter Two).

General Kent's career continued to prosper. In the late 1960s, he was promoted to major general. Subsequently, he was assigned to the Director of Development Planning at Air Force Systems Command (AFSC), headquartered at Andrews Air Force Base, outside Washington. While there, he strove to focus a greater share of the Air Force's development funds on programs intended to "put rubber on the ramp." This meant cutting or canceling outright a large number of long-running efforts that had little promise—a controversial effort but one that bore fruit over time. While at AFSC, General Kent learned a great deal about the hierarchical and inflexible acquisition system within the Air Force and the pernicious effects of the "requirements process," which imposed long and frequently unnecessary delays on the development of new systems and often created controversy late in the development cycle when promising systems failed to meet one or another performance specification. During his time at AFSC, he saw firsthand the problems that arbitrary "requirements" can create—the C-5 transport plane being a prime example as it made its way through various stages of the development process (see Chapter Five).

General Kent's work in OSD and at AFSC had been observed by Gen John McConnell when he was the Vice Chief of Staff. In 1968, when General McConnell was Chief of Staff, he called General Kent to his office and said he was appointing him to head Air Force Studies and Analysis. His explanation for doing so did not touch on General Kent's demonstrated ability to do high-quality analysis. Rather, he said, he needed someone on his staff who would not be a "yes man" but could be counted on for candid advice. General Kent took the job and for the next four years, through the tenure of two chiefs of staff—Generals McConnell and "Jack" Ryan—he was involved in virtually every major issue relating to strategy, operational capabilities, and force structure that the leadership of the Air Force engaged. As General Kent might have expected, McConnell did not confine himself to calling on Glenn to do analysis. He often gave General Kent problems that needed to be handled discreetly. Some sense of the breadth of General Kent's duties under McConnell is communicated by the fact that General McConnell often referred to Glenn as his "junkyard dog."

General Ryan also prized General Kent's extraordinary abilities, although he used somewhat less-colorful language to express his regard for his Director for Studies and Analysis. Under Ryan, General Kent led Air Force Studies and Analysis to its salad days. They developed SABER GRAND, an early theater-level model that was used to generate new insights about the proper apportionment of air assets in a war in Central Europe and that played a key role in the evaluation of a range of modernization programs under Air Force consideration (see Chapter Six). General Kent saw to it that Lt Col Larry Welch (later Air Force Chief of Staff) played a dominant role in defining the performance specifications for the F-15 fighter (Chapter Five). With Jasper Welch, they challenged Navy assertions about the robustness of the submarine leg of the Triad (Chapter Four). Working with Lt Gen Otto Glasser, the Director of Air Force Research and Development, General Kent helped to redirect the justification for the lightweight fighter concept, which begat the F-16 and the F/A-18 (Chapter Five). And following a "Black Friday," on which the C-5, AWACS, and F-15 programs were all deleted from the Senate's Defense Authorization Bill, General Ryan gave General Kent responsibility for developing the rationales for every major Air Force program (Chapter Three).

In time, General Ryan came to rely on General Kent to evaluate virtually every significant proposal and presentation that came to him. General Kent was not reticent about using his own powers of persuasion, as well as his well-known influence with the Chief, to affect decisions about the allocation of the Air Force's resources. In short, when people familiar with the workings of the Air Staff from this period refer to General Kent as having been "the brain of the Air Force," it is not far-fetched.

In 1973, after General Ryan retired, General Kent was named director of the Weapon Systems Evaluation Group (WSEG). The head of WSEG was a three-star general who reported directly to both the Director of Defense Research and Engineering and the Chairman of the Joint Chiefs of Staff. Here, General Kent presided over assessments of the operational utility of many systems—Army, Navy, and Air Force. He was also instrumental in promoting the development of several theater-level combat models, including the Institute for Defense Analyses's (IDA's) Ground-Air Model (IDAGAM), which evolved into the TACWAR model. TACWAR was a blunt but fairly reliable instrument for assessing the outcome of combat in Central Europe, the Korean peninsula, and other theaters. During the 1980s and 1990s, TACWAR was DoD's most widely used campaign model (see Chapter Six). General Kent later came to regard TACWAR and other opaque, theater-level models as rather poor tools, both for estimating the outcomes of operational plans and for informing decisions about the relative merit of various investment options for modernizing operational capabilities. But at the time, sponsoring the development of theater-level campaign models seemed like a good idea.

General Kent retired from active duty in 1974 and became a consultant to Boeing, Northrop-Grumman, and other defense contractors. During these years, he played important roles in a number of programs that have figured prominently in modern U.S. military operations. For example, he helped broker the deal between Boeing and Northrop-Grumman to share development and production of the B-2 bomber. He also gained the support of Gen Robert Dixon, commander of the Air Force's Tactical Air Command, for development of the JSTARS aircraft (Chapter Five).

In 1982, RAND's president, Donald Rice, persuaded General Kent to join RAND as a senior research fellow. He essentially offered Glenn carte blanche to define and undertake research on issues that he felt were relevant to the Air Force and the broader national security community. During his early years at RAND, General Kent's efforts focused on evaluations of the U.S.–Soviet balance with regard to strategic nuclear forces, on ways to enhance NATO's defenses in Central Europe, and on improved approaches to force planning.

President Reagan launched his SDI in 1983,[1] and General Kent published some quite trenchant analyses of that program and its potential implications for national security. His work, which showed that fielding partially effective defenses against ballistic missiles would have deleterious effects on first-strike stability, helped to inject some much-needed realism into an effort that many outside of the administration saw as poorly grounded and misguided (see Chapter Two).

In collaboration with Edward (Ted) Warner and Randall DeValk, General Kent also devised an approach to strategic nuclear arms control that would have given both sides incentives to evolve their forces toward postures that were more survivable and stabilizing as they reduced overall arsenal sizes. An influential group of senators and congressmen embraced the resulting "weapon stations" concept, and, eventually, the Reagan administration's negotiator presented it to the Soviets (Chapter Two).

Ted Warner, David Thaler, and General Kent also collaborated to formalize and refine the "strategies-to-tasks" approach to describing the roles of specific force elements in joint operations. This approach is predicated on a disciplined disaggregation of a campaign strategy, and General Kent showed how it could aid understanding and evaluation of the contributions of specific systems to objectives that were important to success on the battlefield. At the heart of the evaluative mechanism was the notion that every operational task is the "output" of an end-to-end concept of operations incorporating, in a coherent way, the functions of surveillance, assessment, command, control, asset generation, engagement, and attack. By insisting on the importance of each of these functions, strategies to tasks played an important role in highlighting the potential value of a wide range of "information systems"— sensors, software, computer displays, communication networks—that

---

[1]   The goal of SDI, which was more widely referred to at the time as "Star Wars," was, in the President's words, "to make ballistic missiles impotent and obsolete" (Ronald Reagan, "Address to the Nation on Defense and National Security," March 23, 1983).

were just beginning to be integrated into military operations in the late 1980s (Chapter Two).

This volume provides ample evidence that General Kent was remarkably creative and successful in tackling some of the knottiest problems the defense community has faced over the past 50 years. What may not be apparent to the reader are Glenn's qualities as a colleague and friend. He had a reputation, perhaps deserved, of not suffering fools gladly. But no one was more generous with his time and ideas when he believed that you had made an honest effort. Wherever he worked, he developed strong and collegial ties with those who worked with and for him.

Glenn was also an extraordinary mentor to several generations of analysts in the Air Force and at RAND. He would spend hours offering insights about what he regarded as the right way to conceptualize a complex problem, and he would carefully and patiently review draft briefings and reports. In this way, Glenn's first vocation as a teacher and coach remained central to his entire career.

Many view analysis more as window-dressing to support positions previously arrived at than as a way to gain new insights into a problem. General Kent is the antithesis of this cynical persuasion. While he could be a fierce and effective advocate, he often used analysis to change his own service's positions on key issues. This reverence for the art and science of analysis served him and this nation extremely well. When facing a complex issue, he was consistently able to identify key factors and relationships, to create methods that would illustrate how they interacted with one another, and to suggest the most fruitful approach toward a solution. General Kent has remained actively engaged in issues of national security into his nineties, and his mastery of mathematics, encyclopedic experience, quick mind, and work ethic have continued to amaze his colleagues.

General Kent's unique combination of intellectual brilliance, unfailing collegiality, and instinct for the jugular in policy debates made him arguably the premier defense analyst of his era and won him the admiration and devotion of those who worked with and for him. His example should serve to inspire future generations of defense analysts and policymakers as they confront new challenges to the nation's security.

The chapters that follow, all written by General Kent, relate the stories of the roles he played in some of the significant policy issues over which he had influence. The chapters are arranged as follows:

- Chapter One, "The Single Integrated Operationl Plan," first tells the story of the machinations that then-Colonel Kent and the leadership of the Air Force went through in 1960 to bring about the SIOP and the Joint Strategic Targeting and Planning Staff (JSTPS). Following this, it discusses two subsequent events relating to the SIOP and its development.
- Chapter Two, "Nuclear Weapons: Strategy and Arms Control," presents summaries of General Kent's work that broke new conceptual ground on U.S. strategy for posturing and employing nuclear forces, strategic arms control, and first-strike stability between the United States and the Soviet Union during the Cold War.
- Chapter Three, "Analysis, Force Planning, and the Paradigm for Modernizing," relates General Kent's thoughts on how analysis and force planning should and should not be done in the defense community. This chapter also summarizes his approach to running Air Force Studies and Analysis in the 1960s and presents highlights of his approaches to managing and advocating for force modernization.
- Chapter Four, "Modernizing Nuclear Forces," relates General Kent's involvement in a range of U.S. nuclear weapon programs, including the Minuteman missile, manned bombers, the nuclear-tipped Genie rocket, and satellite-based early warning systems, in several cases summarizing analytical techniques that he used to support key decisions relating to these programs.
- Chapter Five, "Modernizing Conventional Forces," does the same thing for General Kent's involvement in conventional forces, including the F-15, F-16, AWACS, and others.
- Chapter Six, "Analytical Tools," discusses a number of tools that General Kent either developed or sponsored, ranging from

campaign-level simulations to fairly simple mathematical constructs.
- Chapter Seven, "Summing Up: Kent's Maxims," offers a distillation of lessons learned over a career spanning seven decades.

A chronology of General Kent's military career and a list of the
awards he has received follow these chapters.

Finally, note that it should be understood that the conversations
General Kent describes in this book are rendered according to the best
of his recollection. Understandably, given the passage of time, precise
verification of quotations would be impossible.

David Ochmanek
Director, Strategy and Doctrine Program
Project AIR FORCE
RAND Corporation

Bruce Pirnie, Ph.D.
Adjunct International Policy Analyst
RAND Corporation

Michael Spirtas, Ph.D.
Policy Analyst
RAND Corporation

CHAPTER ONE

The Single Integrated Operational Plan

Much of General Kent's career was associated, in one way or another, with nuclear weapons. From the early 1950s, he was involved in their development and testing; in equipping the Air Force's inventory of bombers and, later, intercontinental ballistic missiles; in planning for the employment of nuclear weapons; and, by the 1960s, in thinking about how to limit them through bilateral arms control agreements.

Nuclear weapons were at the center of U.S. military strategy throughout the 1950s and early 1960s. Realizing this, the leaders of each of the services were determined that their respective services would "get into the nuclear business" in a big way. The Eisenhower administration's adoption of the New Look strategy formalized the primacy of nuclear weapons. Yet, planning for the employment of U.S. nuclear forces should deterrence fail was not at all integrated. It took until 1958 for the unified and specified commands (including the Pacific, European, and Strategic Air commands) to be created and given operational control over forces assigned to them. Even then, the idea of "joint" or cross-service planning was slow to gain acceptance.

General Kent relates stories of his involvement with U.S. nuclear forces in several chapters of this volume. This chapter shows that he worked on both sides of the centralization issue as it related to planning for the employment of nuclear weapons: He was instrumental in conceiving the idea of a comprehensive, fully integrated operational plan for U.S. nuclear weapons and in creating the organization responsible for developing that plan. Later, he successfully fought to keep the related planning functions in the hands of the uniformed military "in the field" and out of the civilian-

dominated Office of the Secretary of Defense in Washington. In both cases, his approach was first to frame the debate carefully so that decisionmakers would focus on the issues that most strongly supported his case and, second, to show that his preferred outcome would provide best for the nation's security.

The Advent of the SIOP

In the 1950s, I was a colonel heading the Weapons Plans Division on the Air Staff. This division dealt solely with nuclear weapons and related issues. It was subordinate to the Director of Plans, Maj Gen Glen Martin. He, in turn, worked for the Deputy for Plans and Programs of the Air Staff, at the time Lt Gen John Gerhart.

The United States introduced nuclear weapons into its operational plans (OPLANs) in an incremental and less-than-integrated manner. Different entities developed their own OPLANs without much real coordination with others. The Air Force developed a fleet of bombers capable of carrying nuclear weapons. The Navy built submarines that could launch intercontinental ballistic missiles (ICBMs) and also carrier-based aircraft that were nuclear capable. Later, the Air Force created its own array of ICBMs. The incremental nature of these developments resulted in a less-than-integrated plan for how these weapons were to be employed. The United States had no workable approach to unify the planning among the various commands involved.

One day, General Gerhart called me to his office. "There is an ongoing item before the Joint Chiefs of Staff [JCS]," he stated. "The Chief just told me that we cannot afford to lose on this matter. I am hereby making you the 'action officer.' Forget, for a while, about running your division. Focus your entire efforts on winning this action before the Joint Chiefs."[1]

---

[1] For more detail on this debate, see Nathan Twining, Chairman of the Joint Chiefs of Staff, memorandum to the Secretary of Defense, "Target Coordination and Associated Problems," August 17, 1959.

The action before the JCS had been prompted by a memorandum from the Chief of Staff of the Air Force to the JCS recommending that, to have an integrated OPLAN for nuclear strikes on the Soviet Union, the Strategic Air Command (SAC) should assume control of the Navy's Polaris submarines. The Navy and the Army were vigorously and emotionally opposed to this recommendation. SAC was seen (rightly) as an arm of the Air Force, and the Navy, in particular, had a tradition of jealously guarding its autonomy. The fact that the Air Force, the upstart among the services, had grown rapidly under the Eisenhower administration intensified the determination of the other services to prevent this expansion of SAC's authority.[2]

The next day, I discussed the matter further with General Gerhart. I told him that the real issue was whether the United States should have one integrated OPLAN or should just coordinate three plans: one developed by SAC; one developed by the U.S. Navy Commander in Chief, Atlantic Command; and one developed by the North Atlantic Treaty Organization (NATO).

I then pointed out that, to have a single integrated plan, it was not necessary to place the Navy's Polaris submarines under SAC or a unified command. Instead, I recommended that the Air Force advocate the formation of a joint strategic planning group. Personnel from all services would staff this group. The commander of SAC would head the group, and his deputy would be a Navy admiral. I also proposed that this group be collocated with SAC in Omaha, Nebraska.

If we want to prevail, I told General Gerhart, we must abandon the idea of placing Polaris submarines under the control of SAC or even some other specified command, a step that would provoke strong and emotional opposition from the other services. Instead, we should focus on the critical issue of a *single integrated operational plan* (SIOP). We could have a SIOP without grappling with the issue of command and control over nuclear forces. General Gerhart agreed with this approach and recommended to the Chief of Staff of the Air Force, Gen Thomas White, and the commander of SAC, Gen Thomas Power, that the Air

---

[2]   In 1959, the Air Force's share of the DoD budget approached 50 percent. Since 1965, it has never exceeded 39 percent.

Force change its position accordingly. They agreed, though General Power was reluctant, and so the Air Force came to advocate a SIOP and a Joint Strategic Targeting and Planning Staff (JSTPS) in Omaha.[3]

Both the other services opposed our new proposal. In part, this was because the previous attempt to gain operational control had poisoned the well. Even without this, however, the other services would have been suspicious of the proposal because they saw the SIOP and the joint planning staff in Omaha as an ill-disguised gambit by the Air Force to dominate the planning and conduct of nuclear operations, and they were determined to prevent this. In their view, even our new proposal gave the Air Force too prominent a role. They held that it was enough to coordinate plans developed by the services. But we argued that the United States needed a single plan, integrated from the beginning by a joint planning staff.

The Secretary of Defense, Thomas Gates, became aware of this issue and made it known that he favored the approach being proposed by the Air Force. In the interest of advancing the cause of integrated planning, Secretary Gates asked his military assistant, Brig Gen George Brown (an Air Force officer who later became the Chairman of the JCS), to keep him informed of the status of this matter.

In an effort to get the ball rolling in our direction, we proposed a trial run. An ad hoc joint planning staff would convene in Omaha and develop a trial SIOP to test the concept. The Navy opposed this proposal, but the Secretary of Defense took affairs into his own hands and directed the JCS to conduct this trial. After a month or two, the ad hoc planning staff produced a first cut at a SIOP, but there was still considerable dissension between the Air Force and the other two services about a variety of issues.

There was a lot to argue about. The Air Force wanted the plan to take into account the possibility that poor weather would hamper or even prevent operations by carrier-based aircraft. The Navy and the Army pointed to problems regarding several issues having to do with Air Force forces. They argued that far fewer bombers than estimated

---

[3] Thomas White, Air Force Chief of Staff, memorandum to Secretary of Defense Thomas Gates, with attachment on Strategic Targeting Authority, June 10, 1960.

by SAC would be likely to penetrate Soviet air space and reach their targets. They also considered that the Air Force was far too optimistic about the proportion of bombers that could be launched before Soviet ICBMs destroyed them on their bases. In addition, they argued that the Air Force was overestimating the hardness of Soviet industrial targets. They feared that this assumption would have the effect of raising the requirement for more nuclear forces and more nuclear weapons.

In an effort to force a resolution of these matters, the Air Force submitted a memorandum to the JCS, proposing that the Joint Chiefs and the Secretary of Defense convene in Omaha to review the draft plan that the ad hoc planning staff positioned there had developed. Predictably, the Navy and Army opposed this suggestion. Again, the Secretary of Defense intervened. He directed that the meeting be convened on December 18, 1960. In addition, he directed that the objective of the meeting be changed from "review the plan" to "*approve* the plan." The Chief of Naval Operations was so strongly opposed to this wording that he asked to see President Dwight Eisenhower. The outcome of the meeting was predictable: The President sided with his Secretary of Defense.

The ad hoc staff had briefed the plan to the JCS in Washington several times in preparation for the meeting in Omaha. The entire issue was nearing some resolution, and it was certainly a top priority for all concerned. The Air Force intended the following chain of events: The draft plan would be briefed to all concerned (including Secretary Gates) on December 18 in Omaha. The Air Force would then recommend approval of the plan, with expected opposition from the Navy and the Army. The Air Force confidently expected that the secretary would decide in favor of approval, in light of the split, when the decision was put to him. Thus, the plan would be approved.

But General Gerhart wanted nothing left to chance. "How do we know the Secretary of Defense will rule in our favor?" he asked. "We must have a compelling argument to this end. You develop this argument," he told me.

Stuck in traffic as I was going home that night, I pondered how we could make an airtight case for our position. Then it struck me that the right way to think about this was to view the SIOP as an

OPLAN (which it was) and as a means for making the best use of the forces available—with not a hint as to any requirement to meet a set of objectives. The plan should be presented as a set of allocations, i.e., allocating this or that weapon to each designated ground zero (DGZ). What *effects* these strikes would have and whether these were adequate to meet the nation's needs were important but separable questions. One could argue endlessly about such issues as degradation caused by adverse weather and darkness, the probability of our bombers penetrating Soviet airspace, the hardness of the targets, what portion of the Soviets' industrial worth these strikes were likely to destroy, whether these percentages represent enough destruction or too much, and so forth. All these arguments tended to be inconclusive because no one had the experience of all-out nuclear war.

From the perspective of a SIOP—narrowly defined—all these arguments were beside the point. The task given to General Power and his planners was straightforward: Make best use of the forces we have. That meant that there was only one output: the allocation of weapons to DGZs. The Air Force should declare that the sole purpose of an OPLAN was to make optimal use of available forces, which our plan did. There would inevitably be debate about the likely effects of nuclear strikes and whether these strikes were adequate or not, but these were separate issues.

Once this position was staked out, the Air Force needed to show that the protests by the Navy and the Army concerned the likely effects and adequacy of nuclear strikes; they were not about the plan itself. To this end, I proposed that General Power hold a meeting of the ad hoc staff in Omaha. This meeting would include a two-star Navy admiral and a two-star Army general. The participants in that meeting would review all the items that were under protest by any of the services. This would surely be a long and rancorous session. At the end of this session, General Power was to inquire in a quiet manner whether the Navy admiral or the Army general had any recommendations with regard to the assignment of weapons to DGZs. If so, would they please submit those recommendations to him?

We hoped that the other services would make some recommendations, but not too many. General Power would implement all their recommended changes. At an appropriate time during the meeting on December 18, General Power would state that he had asked for recommendations on DGZs, that he had received certain recommendations from the Navy and certain recommendations from the Army, and that he had changed the plan to reflect *all* these recommendations. Therefore, he would declare that the plan itself was not in contest. The likely effects of these nuclear strikes were a topic for continued debate.

General Gerhart liked this gambit. He told the Air Force Chief of Staff, General White, about it and General White called General Power. Without revealing any details over the phone, General White said that he would send me to Omaha to explain the approach. I arrived at Offutt Air Force Base (AFB) late that same evening and was promptly whisked to General Power's office. I explained to him the gambit that I had proposed to General Gerhart. The central point was that the Air Force be able to tell the Secretary of Defense on December 18 that there was no protest about the plan itself.

"It will work," said General Power. "How many others know of this gambit?"

"Only General Gerhart and the Chief," I replied.

"Then keep it that way."

We kept everything under wraps until the meeting on December 18. On that day, all the ranking civilians and military officers were in attendance at SAC Headquarters in Omaha. In fact, General White, a four-star, just made the cut to sit in the first row. The agenda items for the meeting on December 18 all concerned various factors that could change the predicted outcome of the nuclear strikes. As we predicted, the participants engaged in a heated debate about these various factors. Just before lunch, General Power requested the floor. He stood in front of the group and spoke directly to the Secretary of Defense. He explained the difference between factors that affect the outcome of strikes and, on the other hand, the allocation of weapons to DGZs. He said that, with respect to various factors, there was much debate, but that with respect to the allocation of weapons to the DGZs, there was none. He stated that he had accepted all the changes proposed by the Navy admiral and the Army general about the allocation of weapons to DGZs.

Secretary Gates was more than impressed. He said, "General Power, if what you say is true, then this casts quite a different light on this matter." He then extracted a grudging admission from the Navy admiral that he had indeed submitted five changes concerning DGZs and that all had been approved. He extracted a similar admission from the Army general. The secretary then closed the discussion. He said that, since there were no disagreements regarding the allocation of weapons to DGZs, the plan should be approved without further debate, without change, and today.

But the Chief of Naval Operations, ADM Arleigh Burke, was not quite finished. "Mr. Secretary," he said, "I think that it would be rather awkward if the Congress came to know that you coerced the Joint Chiefs of Staff into approving the plan before it was officially submitted to the Joint Chiefs for approval."

Oh my. I had overlooked this one important detail. The plan had been briefed to the JCS three times, and the JCS had a copy of the plan. But the plan had not been officially submitted to the JCS for approval. Any delay would give the Navy time to recover. All the principals were scheduled to depart the premises that afternoon and return to their respective bases. Just as we seemed to have closure, it was slipping away. There was no telling what would happen if the plan were not approved during the session in Omaha. The Navy would recover and now make some stern protest about the allocation to DGZs. I anticipated that General Gerhart would not be pleased with this outcome or with me.

Then Secretary Gates came to the rescue. "Admiral Burke," he said, "you have a point to which I must react. The plan will be submitted to the Joint Chiefs officially today. You will have all night to consider it. I now amend my earlier statement. The Joint Chiefs will report to me in the morning at nine o'clock as to which members approve the plan and which members do not. If one member approves, I expect the matter to be presented to me for adjudication. I will surely find in favor of the member who has voted for approval."

We were back on track.

The Navy admirals and Army generals protested about staying over in Omaha because they had appointments to keep. The secretary brushed the objections aside. "I expect to see each of you in this room tomorrow at nine," he announced. Then he asked, "General Power, may we partake of your hospitality for one more night?"

General Power beamed. "Of course. There will be a reception in the Officer's Club beginning at 1830 hours."

At the reception that night, General Power took me by the arm and ushered me into the presence of General White. "General White, this is the man who made this happen," he said. From that time forward, I was a protégé of General Power, much to the benefit of my career in the Air Force.

The rest is history. In light of the secretary's dramatic statements, the Navy saw no value in continued opposition. At the meeting on December 19, all three members of the JCS voted to approve the plan. Once the decision was made that a JSTPS would be established in Omaha for the purpose of developing a SIOP, the Navy and Army worked hard to make the SIOP a milestone in developing OPLANs.[4]

This SIOP affair was my crowning achievement as a colonel. It is a lesson that big things can be accomplished by diligent and persistent staff work—especially when the Secretary of Defense is on your side. The following underlined my doctrine: Be sure your position in the JCS is so compelling that, if there is a split, the Secretary of Defense will surely rule in your favor. In this case, the key to success was to conceive of a way to frame the debate such that arguments against our position were simply untenable. It was clear from the start that efforts by the Air Force to have the Navy's Polaris fleet "chopped" to the commander of SAC were doomed. But the desired effect—greater coherence in the OPLAN for executing the forces of both services—could be achieved in a different, and more politically palatable, way. Who could argue legitimately against a joint planning staff, especially when it was made clear that the product of that staff—an OPLAN—was intended to make best use of the forces available?

---

[4] Joint Secretariat, "Review of the Initial NSTL and SIOP," note to the Joint Chiefs of Staff on JCS 2056/194, December 9, 1960. For more detail on the process of creating the first SIOP, see Headquarters, Strategic Air Command, "History and Research Division, History of the Joint Strategic Planning Staff: Background and Preparation of SIOP-62," n.d.

In later years, SAC became simply Strategic Command (STRAT-COM), a joint combatant command analogous to U.S. Central Command or U.S. Pacific Command. And more often than not since then, the commander has been a Navy admiral with submarines, bombers, and intercontinental missiles under his control. This arrangement would have been unthinkable to all concerned in the early 1960s. Time changes many things, sometimes for the better.

In this episode, the Air Force gained its point. The result was beneficial not only for the Air Force but also for the country. Had the other services prevailed, the United States would have gone on planning Armageddon in a disjointed way. At best, planning by the individual services would have caused inefficiencies, invited redundancies, and made the nuclear deterrent less credible. At worst, such planning might have caused uncertainty and ragged decisionmaking in a time of crisis. Whatever parochial concerns may have motivated the Air Force to advocate a single integrated plan, it was clearly in the national interest. Finally, one might argue that the SIOP set a standard for jointness that eventually expanded to conventional operations, especially through the Goldwater-Nichols reform.

Defending the Planners of the SIOP

The time was the mid- to late 1960s. Gen John McConnell was Chief of Staff of the Air Force, and Dr. Harold Brown was Secretary of the Air Force. I was head of Air Force Studies and Analysis (AFSA). General McConnell called me to his office. He said that there was a big problem. Three analysts from the Office of the Secretary of Defense (OSD) had made a visit to SAC. They had been briefed on the SIOP and on the "planning factors" used in the preparation of this document. They had then written a report that was very critical of several of the planning factors used by the JSTPS in developing the SIOP.

These analysts had delivered their report to the Secretary of Defense, Robert McNamara. The report included a recommendation that a group be formed to thoroughly review how the JSTPS had developed the various planning factors and how these planning factors were applied to defining the SIOP.

General McConnell went on: "This is not just an effort to improve the SIOP," he said. "These people have a hidden agenda. They have in mind that the 'review group' they are proposing will declare that the JSTPS at Omaha is inept and that this planning should be done by a group reporting directly to the Secretary of Defense." The general added that he had been informed of the existence of this hidden agenda by a very reliable source. He went on to point out that events were moving rapidly. He and Dr. Brown had just returned from a meeting with the Secretary of Defense. Secretary McNamara had informed them that, in view of the report on his desk, he had little choice but to form the group to review the planning. But General McConnell saw this, and rightfully so, as a slippery slope. If the group for the review were formed, there was a strong likelihood that the responsibility for developing the SIOP would be taken away from the JSTPS. He was determined that this whole affair be stopped in its tracks.

General McConnell went on about the discussion between Dr. Brown, the Secretary of Defense, and himself. "While the secretary was explaining that he had no choice," he said, "it suddenly occurred to me that I had a trump card to play. I recommended that the Secretary of Defense not establish the review group until General Kent has had a chance to review the critique by the people from OSD."

The Secretary of Defense accepted this recommendation. He knew me well from my days in OSD under Dr. Brown. The general went on: "Your job is to show that the analysts at Omaha are just as sharp as the analysts from OSD. Show that there are serious flaws in the OSD report."

At that point, Dr. Brown entered the room. "I presume General McConnell has told you of what happened," he said. "I wish to emphasize—there is to be no whitewash. Whatever your findings may be, they must be able to stand up to critical review."

Putting the two statements together, my marching orders were clear: I must show that SAC is right, that the OSD analysts are wrong, and the case must be airtight.

A quick look at the OSD report made it easy to believe General McConnell's statement about the hidden agenda. The critique was wide ranging. It was obvious that the report's authors were trying to establish a basis for taking the responsibility for developing the plan away from the JSTPS in Omaha. They had much more in mind than simply trying to make some improvements to the SIOP itself.

As noted in "The Advent of the SIOP" (pp. 22–30), the SIOP, in effect, defines the allocation of various nuclear weapons among DGZs: Weapon number 22 goes to DGZ number 1; weapon number 23 goes to DGZ number 2; and so on. In rare instances, more than one weapon is assigned to the same DGZ. If you make a change in the SIOP, it is a matter of changing weapon number 45 from DGZ number 21 to DGZ number 28. No big deal. The marginal return of making this change is undoubtedly small and impossible to measure. The effect desired was deterrence: That is, we sought to convince the Soviets that it would be against their interests to launch an attack.[5] In this context, a change in the allocation of particular weapons would result in little change in the degree to which we deterred—especially if the Soviets were unaware of this change, which would be considered Secret or Top Secret.

Now, back to the OSD critique. The analysts identified, as I recall, ten issues that they regarded as evidence of the JSTPS's incompetence. One of these (item 4) caught my eye: "The weight of effort allocated to attack the Tallinn complex is ridiculous." They actually used these words verbatim. This intemperate language stood out and made this item a likely candidate for rebuttal. If I could show that the planners in Omaha were about right in this allocation, we would have a leg up in tarnishing the report by the three OSD analysts.

The Defense Intelligence Agency (DIA) had for some time—up to five years—been reporting on worrisome activities by the Soviets near Tallinn, the capitol of Estonia. Their estimate was that the Soviets were installing an antiballistic missile (ABM) complex to shoot down U.S. missile reentry vehicles (RVs) as they made their way to targets

---

[5] Some people thought the effect desired was to create unacceptable damage in a retaliatory attack. That may be so, but it was always clear to me that we were better served by keeping our focus on deterring a Soviet attack in the first place.

in the Soviet Union. Activities had been observed at some 40 separate sites around Tallinn. DIA was not certain how many interceptors had been deployed (or were to be deployed) at each site and was not all certain of the effectiveness of each interceptor, but it did give a range: between 20- and 80-percent effective—whatever that meant.

The planners in Omaha, in the presence of these tentative assessments by DIA, assumed (1) that 15 interceptors had been (or might be) deployed at each of the 40 suspected sites and (2) that each interceptor had around a 65-percent probability of kill (P_k) given a launch. In the presence of these assumptions, the planners had allocated five RVs per site, for a total of 200 weapons, to suppress the Soviet ABM system. It was the number 200 that bothered the OSD analysts. It seemed like overkill, especially in view of the uncertainty surrounding the complex. Accordingly, they had labeled the allocation "ridiculous."

In truth, at first blush it does seem like overkill (allocating five weapons per site)—if the lethality of each weapon is such that one is all that is required to destroy all the interceptors at one site. But DIA had stated that the probability of intercept might be as high as 80 percent (or words to that effect). The Soviets would use the interceptors at the site in self-defense, and there would be only a 20-percent probability (at worst) of each U.S. RV penetrating, as long as the ABM system is operating. So, it makes some sense to put a sizable number of weapons onto the Tallinn ABM system to ensure that the complex is destroyed and that the U.S. RVs attacking other DGZs are not intercepted. The question remains: What is the optimum number of RVs to commit to attack this complex?

I began to consider ways to quantify the value of suppressing this defense. Starting from the simple case of a single RV against a single ABM site, we can calculate (on an expected-value basis) that the RV would cause one interceptor to be launched in self-defense and destroy 2.8 of the remaining Soviet interceptors: $0.2 \times (15 - 1) = 2.8$. (The 0.2 comes from $1 - 0.8$.) Thus, 11.2 Soviet interceptors would remain. If two RVs were allocated per site, 8.32 Soviet interceptors would remain: $0.8^2 \times (15 - 2) = 8.32$. So there is merit in allocating more than one RV per site. I saw that we should expand these calculations to determine the "optimum number of RVs per site." My measure of *optimum* in this case was obvious: It is the number of RVs used in defense suppression that maximizes the number of RVs that penetrate the defense and proceed to attack other (non–ABM-related) targets on the territory of the Union of Soviet Socialist Republics (USSR).

A discussion with the planners in Omaha revealed that the weight of effort (200 RVs) expended against the Tallinn complex had been discussed—albeit very briefly—with the OSD analysts during their trip to Omaha. The SAC planners were somewhat amazed that the analysts had chosen the word *ridiculous* to characterize this allocation (200 RVs total). The SAC planners pointed out that, if they assumed a probability of intercept of around 65 percent, it took a little more than five weapons per site to attain a damage expectancy (DE) of 0.90 per site. That is, $0.65^5 = 0.12$, and $1 - 0.12 = 0.88$ DE. And that was about the extent of the discussion on that item.

During their visit to SAC, the analysts from OSD did not challenge the number of sites (40) or the possible number of interceptors per site (15), mostly because the subject was not discussed in detail. Rather, they challenged the requirement for 0.90 DE. They opined that while the 0.90 DE might have been sacred to the planners in Omaha, it had no firm basis in policy or mathematical analysis.

I then undertook a simple analysis. The measure of merit was the number of RVs to penetrate the defense and reach DGZs in the USSR. Suppose that 1,000 RVs were involved in an attack. The planner would be willing to divert 200 of these RVs to defense suppression if, by doing so, the number of RVs available to attack other DGZs would increase.

The notional characteristics of the Tallinn ABM complex were as follows:

1.   40 surface-to-air missile (SAM) sites
2.   15 interceptors per site
3.   for a total of 600 interceptors.

Table 1.1 reveals that, for the conditions stated, the optimum number of RVs per site in defense suppression is between four and five. This allocation maximizes the number of U.S. RVs that penetrate the Soviet defense, as shown in the far right column of the table.

Table 1.1
Assessment of Allocation Options: RVs to Suppress ABM Defenses

RVs per Site (no.)	Defense Suppression RVs (no.)	ABM Interceptors			U.S. RVs Destroyed (no.)[a]	RVs That Penetrate (no.)[b]
		Not Fired (no.)	Surviving (%)	Remaining (no.)		
0	0	600	100	600	480	520
1	40	560	80	448	358	602
2	80	520	64	333	266	654
3	120	480	51	245	196	684
4	160	440	41	180	144	696
5	200	400	33	132	106	694

[a] For a probability of intercept (PI) of 0.8.

[b] The number of RVs launched less the number expended in defense suppression less the number destroyed equals the number of RVs that eventually penetrate the defenses.

So it turns out, fortuitously, that the planning factor of 0.90 DE and a probability of intercept of 0.65, as used by the SIOP planners, gave approximately the right answer, though perhaps not for entirely compelling reasons. But I was not obliged to defend the reason. The OSD analysts in their report to the Secretary of Defense had not directly challenged the use of a DE of 0.90. Rather, they had challenged only the number of RVs diverted. So with regard to item 4, OSD was wrong, and the planners in Omaha were right. I explained all this to General McConnell and, in turn, to Dr. Brown. Dr. Brown went through my calculations number by number and was satisfied that the logic was sound.

Dr. Brown then convened a meeting with the general, with the three OSD analysts who had written the report and myself in attendance. I started with a dissertation on the strategy to be followed when planning in the face of uncertainty. I noted that planners must make choices about the allocation of forces and that their choices must be predicated on judgments about a wide range of factors whose precise values are unknown. To be robust, a plan must be executable in the face of adverse circumstances. Hence, the correct strategy is to maximize the outcome for the case in which the factors in question are adverse. We should, in the vernacular, "maximize the min." We should plan against the worst case. In this case, that meant planning for the case in which the probability of intercept (PI) for the Soviet ABM system was 80 percent. It would be imprudent to adopt a planning factor that maximized the outcome when the PI was 20 percent.

Notice that I had reduced the problem to whether the PI to use was 20 percent or 80 percent. This spread was derived from the DIA estimates. I announced then that allocating five RVs per site, a total of 200 RVs against 40 sites, was about right for the factor of 0.80. I added that simple math could be used to demonstrate this.

I had expected that rather than challenge my math, the analysts from OSD would challenge the assumptions about the structure of the complex I had assumed—especially that the number of interceptors per site was 15. If one assumed that fewer than 15 interceptors were present, the optimum number of RVs per site would be less than five. To my surprise, they chose not to challenge either my math or my assumptions. Instead, they declared that they were "not going to get in a numbers game" with me.

I replied, "You raised the issue of numbers. You stated that an allocation of 200 RVs (five per site and 40 sites) was 'ridiculous.' My math demonstrates that five is about right. Where is your analysis that shows that five is so far wrong that it warrants the characterization of 'ridiculous'?"

At that point, Dr. Brown observed that my confrontational tone was not conducive to a productive discussion, whereupon General McConnell, to my surprise and chagrin, stated that he agreed with the secretary and summarily dismissed me from the room. As usual, the general was one step ahead. In about 30 minutes, he called me to his office. "I got you out of the room," he said, "before they could change tactics and try to challenge your analysis. After you left, I made the point that, if they continued to pursue this matter, I would make the point that they, when faced with your analysis, refused to be drawn into a numbers game. I would certainly make this known to the Secretary of Defense—and anyone else who would listen." The general went on: "While they did not agree to cease and desist, I think we have heard the last of this matter. They, of all people, should avoid making sweeping statements before they do their homework."

And so it was. Their report withered on the vine. They no longer pushed the matter. Their choice of the word *ridiculous* was ill advised, and I had turned it into a fatal error.

In summary,

- In cases of this nature, it is almost always preferable to attack than to defend. It would have been very difficult for me to show convincingly that everything the JSTPS was doing was correct. At best, this approach would have resulted in a series of charges and countercharges with no conclusive resolution. But if I could discredit the report of the OSD team, we might convince Dr. Brown and Secretary McNamara to drop the affair.
- If you undertake to challenge or discredit a report, focus and dwell on the one item about which its authors are obviously wrong and be prepared to prove your point.
- Never use such polemical words as *ridiculous* in your own work unless the number you are challenging is very wrong and you can prove it. Even under those circumstances, find a less inflammatory word.

In the wake of this episode, I was elevated to a higher stature in General McConnell's eyes. He would at times introduce me as his "junkyard dog."

Calculating the "SIOP Degrade"

In 1991, Maj Gen Robert Linhard was the Director of Plans and Resources at SAC in Omaha. I had retired from the Air Force and was working at the RAND Corporation. There had been some discussion in Washington about deploying an active defense to counter any attack against the United States by a third country (that is, one other than the Soviet Union) with nuclear-armed ballistic missiles. Such a deployment would not be consistent with the ABM treaty of 1972. Rather than withdraw from the treaty, the United States was considering a concept that would allow the Soviets to deploy a similar defense along their southern borders to guard against an attack by their neighbors to the south.

General Linhard wanted to know how such a defense, if allowed, would affect the United States' ability to execute the SIOP. In the terms he used, he sought to understand the "SIOP degrade." To this end, he invited a group of analysts to convene in Omaha to address this question. The invitees included analysts from the Los Alamos, Livermore, and Sandia National Laboratories; two analysts from the RAND Corporation (Dean Wilkening and myself); one representative from the U.S. Arms Control and Disarmament Agency (ACDA); someone from the Department of Defense (DoD) Office of Research and Engineering (DDR&E); and others.

The meeting opened promptly at 0830. General Linhard stated his question and problem. The three analysts from the national labs set forth their approach to answering the question. After further questions and discussions, we finally disbanded for lunch. After lunch, General Linhard opened with a statement: "What I learned this morning is that, if I gave each of you a million dollars, after six months of 'computer crunching,' you could provide some tentative answers." They all nodded affirmatively.

At two o'clock it was my turn. I opened by saying, "General Linhard, I am prepared to provide you considerable insight as to the implications of such a Soviet defense, this afternoon and for free. First, the bottom line: The degrade to the SIOP will be minimal (something like 10 percent), provided (1) that you employ decoys based on the latest decoy technology and (2) that you forbid the Soviets to deploy any forward engagement radar in the northern part of their country (a radar to control the engagement of a Soviet interceptor engaging U.S. RVs)."

To scope the problem,

1. Assume that each side is allowed no more than 200 interceptors.

2. Assume that each interceptor has a P_k (for a given engagement) of 0.7, for a total kill potential of 140.

3. Assume that the attack by Blue is 1,000 RVs against 1,000 targets, one RV per target.

4. Assume that 1,000 targets contain people and industrial facilities amounting to 1,000 units of "worth"—the proper choice of a scaling factor can make this true. Note that, for convenience, one unit of worth is an erdel (*erdel* is a fabricated word and means nothing).

5. Assume that the distribution of worth (erdels) among the 1,000 targets obeys the distribution according to the equation by the renowned economist Vilfredo Pareto, in which

$$V_{cum} = \left(\frac{n}{N}\right)^{\frac{1}{2}} \times w,$$

where V_{cum} is the cumulative value, n is the stated number of targets, N is the total number of targets in the set, and w is the worth in the total set. The 1/2 is the exponent Pareto derived, which of course indicates the square root. It follows that one-half of the total value is in the first one-fourth of the targets.

Note that I was not breaking new ground with regard to the distribution according to Pareto. Many analysts have used this distribution, including the assignment of one-half as the value of the exponent.

6. Assume that the Soviets are not allowed to deploy an "engagement" radar to the north of the tier of provinces in the south of their empire.

7. Without an engagement radar to the north, they cannot engage in "threat-tube sorting." That is, they cannot determine the intended target DGZ of each attacking RV and thus employ their 200 interceptors against the 200 RVs that are destined for the 200 most lucrative targets. Note that the defense has great leverage if it has a control system that provides this capability. Without it, the defender has to engage the incoming RVs with-

out regard to the value of the target that each RV is intended to destroy.

8. Also, by not allowing the Soviets to deploy an engagement radar to the north, we assume that they cannot discriminate between decoys and RVs in the attack by Blue.

For the base case (no defense), the defense of course saves no worth (zero erdels), and there is no degrade to the SIOP. Now take the case of a defense with a kill potential of 140 (200 × 0.7). There are 1,000 RVs versus 1,000 targets, and each target contains (on the average) one erdel of worth. Thus, the defense "saves" 140 × 1.0, or 140 erdels. This amounts to a 14-percent degrade.

For the next case, add 400 decoys to the attack. The decoys are, perhaps, Mylar balloons and very light. Assume that we must allocate 20 weapon spaces to employ the 400 decoys—a rate of 20 to one. The 980 remaining weapons attack 980 targets. These 980 targets contain 990 erdels:

$$1,000 \times \sqrt{\frac{980}{1,000}} = 990.$$

Now, with 400 decoys (and 1,380 total objects), the expected erdels saved per object killed by the defense has been reduced from 1.0 to 0.717:

$$\frac{990}{1,380} = 0.717,$$

assuming that the Soviet defense system is unable to distinguish between RVs and decoys. Then, the erdels saved by killing 140 objects is equal to 100. To this number, we must add the erdels saved by replacing 20 weapons with decoys. This number is 10. So, the total erdels saved by the defense is 110, and the degrade is 11 percent. If the technology of lightweight decoys permits 40 decoys per weapon space allocated rather than 20, the degrade will be a little less. The defense now "saves" less per object killed, 0.556 versus 0.717, as was the case for 400 decoys. Now, the defense saves 78 erdels (140 × 0.56)—plus the 10 erdels in the 20 targets not attacked—for a total of 88 erdels, a degrade of slightly under 9 percent for the case of 800 decoys.

On the other hand, if the Soviets can do threat-tube sorting, the problem takes on a different complexion. With threat-tube sorting (and no credible decoys), the Soviets would allocate their 200 interceptors to attack (selectively) the 200 RVs that are directed toward the 200 most lucrative targets. These 200 RVs put at risk nearly 45 percent of the total value of the system of 1,000 targets. The expected value saved by the defense per RV destroyed is now 2.25, and a kill potential of 140 saves 315 erdels. Now, the degrade is a whopping 32 percent. We can see the powerful reasons for outlawing forward engagement radars: (1) There is no threat-tube sorting, and (2) there is no discrimination between RVs and decoys.

At this point, some of the other analysts could see their prospects for substantial future work slipping away and proclaimed that the problem could not reliably be reduced to such a simple calculus. Both General Linhard and the gentleman from ACDA thought otherwise, and they stated that they now had adequate insight to inform some key policy decisions:

1. Outlaw forward engagement radar (or radars).
2. Hold the number of interceptors the Soviets were allowed to a low number—a few hundred.
3. Pursue a vigorous program to develop lightweight decoys.

In the presence of these three provisos, the problem of SIOP degrade was minimal and presented no compelling argument against negotiating with the Soviets to allow both countries to deploy a limited defense to counter the threat of attack by third countries.

The key here is to scope the problem and address the key assumptions in a manner distinct from a focus on "computer crunching." In other words, just sit back and think. Often, doing this can provide a basis for calculations that, while quite straightforward, yield new insights into the most important aspects of the problem.

For reasons external to those examined here, the whole concept of each side deploying a limited defense was lost in the turmoil surrounding the collapse of the Soviet Union and the United States' decision, in 2001, to withdraw from the ABM treaty of 1972.

CHAPTER TWO

Nuclear Weapons: Strategy and Arms Control

Over a period of more than 25 years—from the early 1960s through the late 1980s—General Kent conducted what was, perhaps, the most original, insightful, and rigorous work on nuclear strategy and arms control in the nation. In some cases (notably, in his work on limiting damage), his research decisively shaped U.S. policy and resource allocation. In others (as in defining the conceptual framework for strategic nuclear arms control), for reasons having nothing to do with the relevance or quality of his work, he was not entirely successful.

As the cases described in this chapter show, General Kent's work on these issues always focused on finding ways to advance one or more explicitly stated national objectives—limiting damage from a Soviet nuclear attack, defining arms control regimes that would enhance stability, determining the effects that deploying strategic defenses would have on first-strike stability. Equally important, he devised analytical approaches to the problems at hand that were logically and quantitatively rigorous yet transparent to decisionmakers. These qualities gave his work a cogency that is all too often lacking in the realm of defense analysis.

Limiting Damage to the United States

Early in 1964, a study group was formed under the auspices of OSD's Office of Systems Analysis, headed by Dr. Alain Enthoven, to "look at" a wide range of issues relating to U.S. nuclear forces and strategy. Dr. Frank Trinkl was to run this group for Dr. Enthoven. At the time, I was a brigadier general working in DDR&E for Dr. Harold Brown.

As someone with experience in nuclear forces, I was asked to serve on this group. I declined. In my view, there was next to no chance that this group would succeed in providing insights useful for informing decisions on the critical issues of the day. Several things were wrong:

1. There were far too many issues to address (18 in all).
2. There were far too many people involved. Some had been called to the Pentagon on temporary duty and had no place to work.
3. There was no hint as to the analytical approach that would be used to gain insight on any particular issue.

That I had declined to serve was made known to Dr. Brown, my boss. He called me into his office. I told him of my misgivings about the way the effort was being run. He rather agreed but directed that I serve anyway and do my best to make it a useful endeavor. To me, this amounted to a mandate to try to reshape the effort.

My first recommendation had to do with "grouping." There were several issues on the study group's agenda that had something to do with limiting the damage to the United States that might be caused by nuclear attacks. Why not bundle them together under one large effort called "limiting damage" and conduct a comprehensive analysis of the prospects of significantly limiting the damage to the United States from a Soviet attack and, additionally, define the proper allocation of resources to the various means for doing so? These means would include the following:

1. counterforce operations
2. active defense against ballistic missiles
3. active defense against bombers
4. passive measures (civil defense).

Dr. Trinkl opined that such a study would require a major effort that he did not have the time to undertake. I pointed out that I had not introduced any new or additional items. I had merely laid out an analytical approach to gain insight into a major and encompassing issue. Indeed, one of the items he listed was just that: "limiting damage." Absent a comprehensive study, we were doing nothing more than "looking at" several items without providing much-needed insight on any one of them.

Dr. Trinkl repeated that he was reluctant, given his time constraints, to undertake the substantial effort I had proposed. He had ample reason for this reluctance. Dr. Trinkl held fast, and so I departed the group. I made sure that Dr. Brown heard of my decision (and of my reasons for it) from me and not from someone else. He listened and nodded. "You are right. Do the study yourself," he said. Ouch. I was now embarking on an effort that would consume most of my time and energy for the next year or so.

The first thing I did was recruit some good people. My choice from the Air Force was Maj Jasper Welch. I knew him from earlier days. I asked a Navy admiral I knew for the name of a Navy officer to serve on my team for this huge effort. He mentioned a Captain Paolucci. "But you don't want him. He is something of a maverick. He is very smart but can be difficult to work with."

I thought for a moment. "Send him down," I said. "I would like to talk to him."

Capt Dominic Paolucci opened the interview by interviewing me. "What am I getting into?" he asked. "I don't want to waste my time working for another flag officer who doesn't know squat about analysis." I explained my approach and showed him some notional plots. Also, I added, we would try to provide insight as to the allocation among the various means (players) based on marginal return.

Captain Paolucci was all ears and he began to get interested. "I would like to serve," he said. Recruiting Captain Paolucci was one of the best moves I made. He was a brilliant analyst and became one of my best friends.

After several weeks, Major Welch, Captain Paolucci, and Captain Niemela of the Army, using mostly notional data, produced a first cut, "proof of principle" assessment of the primary options for limiting damage from a Soviet nuclear attack. We presented this preliminary briefing to Dr. Brown as an update about work in progress. To our surprise, he then showed these charts to Secretary McNamara, who in turn directed that there be a comprehensive effort with participation by each service and, as well, the staff of the Civil Defense Office.

Suddenly, the stature of the study went up dramatically. This was both good news and bad news: The representative from each service was to be a two-star general; I was only a one-star. There was also an effort to establish a "steering group," ostensibly to advise my group on the conduct of the study. I deftly avoided such a group. We were now engaged in a comprehensive study that was to have many twists and turns.

One of the first things to do was to get a better cut at the database for population and manufacturing value added (MVA). A contractor laboriously went through U.S. census data and defined circles in rank order as to their "worth," according to population and their value in terms of MVA. The location of the first two circles was not surprising: They were in Manhattan. Only one circle in California made the first ten. At that time, California did not account for as large a portion of the U.S. population or gross domestic product as it does today. Moreover, even though the state ranked high in population, it was quite dispersed. The circles, as I recall, had a radius of five miles— approximately the lethal radius of a 1-megaton bomb.

So that was our construct. The United States would strive to defend the worth in each of these circles, and the Soviets would attempt to destroy it. We assumed that the Soviets would attack where the worth destroyed per missile expended would be greatest. These are the most lucrative DGZs. Needless to say, this database became close-hold. The circles were ordered in the first instance by population.

We also had to consider the fatalities among the population from the radioactive fallout from Soviet counterforce attacks on our nuclear forces, the Minuteman silos in the Dakotas being the prime example. If the wind came from the northwest, the debris from these attacks would cause considerable fallout over Chicago and surrounding areas. To determine the number of fatalities from this fallout, we had to make assumptions about the proportion of the population that would seek refuge in fallout shelters, the proportion that would survive if they did, and the proportion that would survive without shelters.

We addressed the Soviet bombers' attack as follows: We would deploy enough active air defense in sufficient numbers of interceptors so that the Soviets would see no advantage in acquiring and deploying bombers as part of their attacking force.

Obviously, there were many assumptions to discuss (and argue) with the participants from the services, each a two-star with a "dog in the fight." Finally, we had some answers. We were careful to state that the study was not intended to predict the outcome of an attack with any precision. Rather, it was intended to provide insight into two issues:

1.   What are the prospects for limiting damage to the United States from a determined and adaptive Soviet attack?[1]
2.   What is the proper (best) allocation of resources among the various "players" involved in limiting damage:
    a.   civil defense with passive measures, such as fallout shelters in many areas and blast shelters in the largest cities
    b.   active defenses against arriving Soviet ballistic missiles with Nike-X
    c.   counterforce attacks with Minuteman missiles against Soviet ICBM bases, submarine-launched ballistic missile (SLBM) ports, and bomber bases
    d.   antisubmarine warfare operations with U.S. submarines against Soviet SLBMs at sea
    e.   active defense with U.S. interceptors versus Soviet bombers?

We arrived at these allocations based on a marginal return.[2] We finally published the study. The essence of the study is detailed in Figure 2.1. The figure shows, for each utility level,

1.   the total resources the United States must spend to achieve that utility level in the presence of the stated Soviet attack

---

[1]   We were careful to construct the analysis so that the Soviet attack could be adjusted to whatever defensive measures the United States put in place. Failing to do this would have led to results that overstated the value of the defenses.

[2]   Unfortunately, I have misplaced many of these graphs over the years, and they are too complicated to reproduce now.

Figure 2.1
Typical Allocations of U.S. Damage-Limiting Forces: Soviet Second-Strike Countervalue

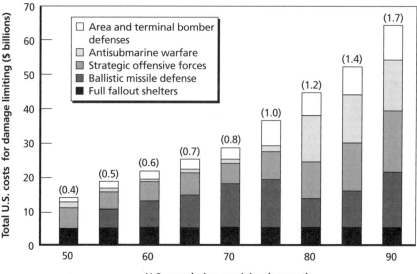

SOURCE: Directorate of Defense Research and Engineering, *A Summary Study of Strategic Offensive and Defensive Forces of the U.S. and USSR*, Washington, D.C., September 8, 1964, p. 120.

NOTES: For each bar, the number in parentheses is the approximate ratio between the cost of U.S. damage limiting and the cost of Soviet damage creation.

In this example, the Soviets allocated $12 billion for ICBMs, $16 billion for SLBMs, and $9 billon for bombers, for a total of $37 billion (FY 1965 cost estimates).

RAND OP223-2.1

2.   the appropriate allocation of resources to each player
   a.   civil defense
   b.   ballistic missile defense (BMD)
   c.   strategic offensive forces
   d.   antisubmarine warfare
   e.   bomber defense
3.   the ratio between U.S. expenditures to limit damage and Soviet expenditures to create damage as required to maintain the stated utility level.

For example, the figure shows that, to achieve the level of 70-percent survival of the U.S. population (against a stated Soviet deployment of ICBMs and SLBMs), the United States would have to spend a total of $28 billion. If the Soviets reacted to our deployments to limit damage by deploying more ICBMs and SLBMs, the United States would be obliged to spend more money to stay at the 70-percent level. The exchange ratio (the amount the United States would have to spend to limit damage compared to the amount the Soviets would have to spend to create damage) was adverse to the United States. At the margin, it was always cheaper to create damage than to limit damage. The graph shows exchange ratios of 0.8 and 1.7, respectively, at the 70-percent and 90-percent survival levels. We realized (belatedly) that the published numbers, which reflected the official exchange rate between the ruble and the dollar, understated the exchange ratio. When the values were revised on the basis that the costs to the Soviets to purchase ICBMs and SLBMs were comparable to our own costs, the ratio was more like 2:1 at the 70-percent survival level. At the 90-percent level, the ratio was more adverse—probably 6:1. The charts actually briefed to Secretary McNamara reflected the revised figures.

The secretary observed that this was a race that we probably would not win and should avoid. He noted that it would be difficult indeed to stay the course with a strategy that aimed to limit damage. The detractors would proclaim that, with 70 percent surviving, there would be upwards of 60 million dead.

The secretary went on: Instead of seeking unilaterally to limit damage, we should undertake to negotiate a treaty with the Soviet Union to curtail the deployment of nationwide defenses. This could set the stage for agreements to control the deployment of offensive forces. Needless to say, this was a statement of great and lasting strategic importance.

So my efforts to gain insight into options for limiting damage had an unexpected ending. Rather than reordering priorities for investments among approaches to limiting damage, the study resulted in a rather fundamental change in policy that led the administration to more or less abandon efforts to limit damage in a meaningful way.

"Limiting damage" did not appear as a stated strategic objective in the draft presidential memorandum (DPM) issued in 1964.

Limiting Damage: Allocation of Resources

As stated earlier, the study on limiting damage addressed the question of the proper allocation of resources among the various "players":

1. civil defense
2. BMD
3. strategic offensive forces to engage in counterforce operations against Soviet ICBMs in silos and Soviet SLBMs in port
4. antisubmarine warfare forces to conduct counterforce operations against Soviet submarines at sea
5. air defense forces (both area and terminal defenses) to intercept Soviet bombers.

The allocations, in general, were based on marginal return. All the players operate with diminishing returns, and we have the optimum allocation when all players are operating with the same marginal return, that is, when a stated increment of funds yields the same increase in the measure of merit (U.S. population surviving) regardless of the player to which this increment is granted.

We determined the optimum allocations using graphs. We plotted graphs with the measure of merit on the ordinate and money allocated to a particular player on the abscissa. Once the graph was plotted, we could take a ruler and determine the slope of the line at any particular level of expenditures allocated to that particular player. The slope, by definition, is the ratio of the change in the measure of merit to the change in investment, where the measure of merit was either the population surviving or the MVA surviving (see Figure 2.2).

We created such a plot for each of the five players. It was, to say the least, a complex, iterative, and laborious approach. But the team persevered, and finally, we arrived at a point at which, with reasonable confidence, we could state the optimum allocation of resources among the various players.

Figure 2.2
Relationship Between Investments in BMD and U.S. Population Surviving (notional)

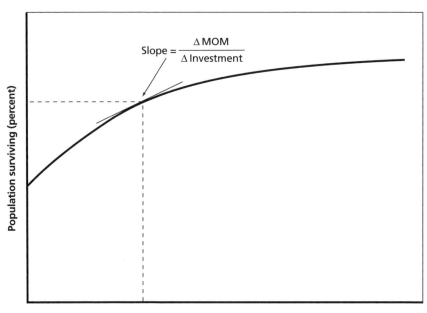

Investments in BMD ($ billion)

For example, we concluded that, for the stated size of attack, the United States had to spend a total of $28 billion for 70 percent of the population to survive. That $28 billion should be allocated as follows:

- $5.2 billion to civil defense
- $12.8 billion to active BMD
- approximately $6 billion to strategic offensive forces
- approximately $1 billion to antisubmarine warfare
- approximately $3 billion for bomber defense.

Today, with high-speed computers, the optimum allocation to each player (at various levels of total expenditures) could be determined more quickly but not necessarily more reasonably. We conducted many excursions to test the proposition that allocations other than the ones we stated as optimum would yield worse results in terms of the measure of merit (U.S. population surviving).

One challenge of note regarding allocations stemmed from a group of people on contract at the Civil Defense Agency. They were charged with evaluating the concept of constructing blast shelters in the largest cities. We chose not to implement this concept even though, at the higher level of expenditures on BMD, the marginal return was equal to or better than the slope for BMD for population surviving. What disqualified blast shelters from our final comparisons was that, unlike every other player, they offered no protection to the economic infrastructure of the United States. In short, when the measure of merit is MVA surviving, blast shelters for the people were not considered a player. Since BMD operates on both of the measures of merit, the nod goes to BMD. Not surprisingly, the Civil Defense Agency contractors were not convinced and continued to challenge our decision not to recommend that this concept be implemented. As it turned out, none of the concepts were implemented. In this sense, one could say that the operation (the analysis) was a success, but the patient (limiting damage) died.

Another challenge regarding allocations came from the representative that the Air Force assigned to my study group. We had stated that, in each of our optimum allocations, the United States should provide the capability to employ one effective RV against each Soviet ICBM silo. This was an increase from the planning factor of zero set forth in the DPM, developed by Dr. Enthoven. This was not a new issue. Dr. Enthoven argued, with some traction, that the United States, as a matter of policy, would not engage in a first strike. Thus, in any conflict, most of the Soviet ICBMs would already have been launched before our RVs arrived. This meant that our counterforce attacks against these empty silos would have minimal effect in reducing the number of Soviet ICBMs arriving at U.S. targets.

Our analysis on limiting damage revealed that counterforce operations with Minuteman RVs (one Minuteman per Soviet silo) were a viable option, even if only 20 percent of those silos still housed an ICBM when the Minuteman missiles arrived. We stated that we could not predict with any certainty how a future exchange might unfold.

Still, in the presence of this uncertainty, we should adopt a planning factor of one RV per silo.

Dr. Brown and Dr. Enthoven were impressed by this argument, and it was agreed that DoD would revise the planning factor up from zero to one. But the Air Force representative, a two-star general, wanted more. He argued for a planning factor of two RVs per Soviet silo. To bolster his position, he argued that the P_k of a Minuteman RV against a Soviet silo was 0.9. We had used 0.6 for the P_k, and with good reason.

I pointed out to the Air Force general that the marginal return of the second Minuteman RV would be minimal, as the P_k of the first RV was 0.9. This was so because 0.9 times 0.1 (the residual value of the silo that had already received an RV) is 0.09. If you choose to argue for two, I said, then you should allege that the P_k is 0.5, since x times $1 - x$ is at a maximum when $x = 0.5$. Still, the general held to the position that, if we would just use 0.9, the stage would be set for a planning factor of two, when in fact alleging that the P_k was 0.9 destroyed any argument for the second weapon. As a matter of fact, a P_k of 0.6 was not far off from 0.5 in terms of the marginal return of the second weapon: It was 0.25 for a P_k of 0.5 and 0.24 for a P_k of 0.6. Nevertheless, the Air Force general refused to grasp the disconnect in his train of logic.

He had other complaints as well. He bundled them together and proclaimed to the Chief of Staff of the Air Force (then Gen Curtis LeMay) that I was "selling the Air Force down the river."

General LeMay convened a meeting, the stated purpose of which was to determine whether or not I was selling the Air Force down the river. General LeMay began the meeting by stating his intention to demote me back to colonel. At that meeting, the two-star presented his case—including the issue that he had submitted that the P_k was 0.9 and I had ignored his input and used 0.6. In my response, I dwelt on this point. Fortunately, General LeMay saw the disconnect in the logic presented by the two-star, namely that a P_k of 0.9 did not set the stage for a planning factor of two. In fact, it was the other way around. I pointed out that I had convinced Dr. Brown and Dr. Enthoven of the merit of allocating one Minuteman RV to each Soviet ICBM silo—up from zero. I pointed out that the issue of using 0.6 for the P_k as a planning factor was hardly any evidence that I was selling the Air Force down the river. As a matter of fact, it was quite the contrary. At this point, General LeMay abruptly ended the discussion and departed the room. The crisis abated, and I heard no further word on this matter.

Helping with DPMs

During the tenure of Secretary Robert McNamara, there was, as I mentioned above, a document known as the "draft presidential memorandum." Actually, it was none of these things. It was not a draft; it did not go to the President; and it was not a memorandum. Nevertheless, it was a defining document whose focus was on defense strategy and the associated means. It defined objectives and the strategy we would use to achieve them. It also discussed the means and capabilities we intended to use to implement our strategy and the forces we intended to field to gain these operational capabilities—all in considerable detail. It also contained programming data, such as the amount of money allocated to each program element in the budget to field these forces.

All the above was administered by the Planning, Programming, and Budgeting System (PPBS), then run by Dr. Charles Hitch, formerly of RAND. Dr. Enthoven, also formerly of RAND and who reported to Dr. Hitch, was responsible for writing the portion of the DPM that dealt with strategic nuclear forces and strategy. The overall construct was a marvel of clear thinking. The section by Dr. Enthoven was invariably logical, insightful, correct, and short—on the order of 30 pages. I was heavily involved in the preparation of the strategic nuclear section simply because of Dr. Brown. Dr. Hitch had to gain approval from Dr. Brown, and Dr. Brown consulted me on what was to be stated in this section.

I address the matter of the DPM not to chronicle my part (which was secondary) but rather to compare this document and to contrast its development with the approach used today. Documents in the mid-1990s, such as the *Report on the Bottom-Up Review* and the Secretary of Defense's annual reports, compare favorably. But not so today; the Strategic Planning Guidance recently issued is 300 pages long, and it fails, in my judgment, to define in clear terms either our strategy or the means we intend to use to implement that strategy.

Dr. Hitch, Dr. Brown, and the secretary himself would review the DPM, item by item and line by line. In addition, the Chairman of the JCS and the service chiefs conducted their own review. Their comments were taken seriously, but they held no power of veto.

One example in particular illustrates the clarity of the document. The document stated the administration's paramount strategic objectives: to deter strategic nuclear attack by the Soviets on the United States; to limit damage from such an attack if one should take place; to deter the Soviets from launching an all-out attack on Western Europe; and, if necessary, to halt such an invasion as far forward as possible (i.e., near the intra-German border).

Our strategy with respect to the first objective, deterring the Soviets, was to have the operational capability to inflict severe damage on the USSR. We gained this capability by deploying SLBMs on submarines at sea, by deploying weapons on bombers on quick alert at bases in the United States, and by deploying nuclear weapons on ICBMs in hardened silos in the United States.

Prior to the completion of the damage-limiting study, our strategy with respect to limiting damage was to increase our operational capability to conduct counterforce operations against Soviet ICBMs in silos and SLBMs on submarines in ports, to deploy an active defense against incoming Soviet RVs and bombers, and to construct a stated number of fallout shelters (but no blast shelters). From this point, having laid out the basic defense strategy, the DPM dealt mostly with programming data.

The DPM was indeed a defining document in shaping the direction the United States was to take—both with respect to strategy and with respect to means. The DPM reflected the decisions of the time and was revised annually until it was eliminated by Secretary Melvin Laird during the Nixon administration. The DPM of 1963 included some definitive guidance as to measures the United States would take to limit damage. In the DPM of 1964, these statements were quietly absent, a change stemming from the study on damage-limiting I had conducted.[3] In light of that study, Secretary McNamara had decided against allocating large amounts of resources to the various means of achieving that objective.

Over the years, I had, from time to time, several arguments with Dr. Enthoven, some of which were quite heated. On the other hand, I always applauded his skill and insight in drafting the DPMs. We would do well to emulate his approach and construct today.

Changing the Paradigm of Arms Control

I first became interested in arms control when I was a fellow at the Center for International Affairs at Harvard, beginning in 1961. At the time, there was a joint arms control seminar populated by interested faculty and selected graduate students from Harvard University and the Massachusetts Institute of Technology (MIT). I was privileged to attend the seminar through the auspices of Dr. Thomas Schelling, a distinguished professor at Harvard and later a Nobel laureate.

Two of the professors involved in the seminar, one from Harvard and one from MIT, had been commissioned by ACDA to define the parameters of a possible agreement to limit nuclear weapons that might be used as a basis for negotiations with the Soviets. These professors presented their work at a meeting of the joint seminar.

Their proposal centered on creating a regime whereby each side would agree to deploy mobile ICBMs in a defined and quite constrained deployment area. Participants in the seminar offered a number of comments at that session, most of which focused on questions of verification. At the conclusion of the meeting, the powers that be decided to devote the following meeting to a more in-depth discussion of the proposal and invited the participants to examine it in greater detail.

At the next meeting, I offered a brief summary of an analysis I had done of the proposal. My work showed that the proposal could have negative implications for the survivability of the retaliatory forces of the adversaries. More specifically, I found that under the terms of the

---

3   For more detail, see "Limiting Damage to the United States," pp. 53–50.

proposed agreement, one side or the other might be able to achieve the capability for a robust second strike after incurring a first strike by the other, but it was not possible for both to do so.

I constructed a table that showed the number of missiles that side A must deploy, A_{req}, in the constrained area to ensure a second-strike capability, given that side B has deployed a certain number of missiles, B_{dep}. In the analysis, I assumed that each side would wish to assure itself that 300 missiles would survive a first strike by the other. The inputs were as follows:

1. the size of the constrained deployment area (the proposal had designated an area of 50 km by 50 km—2,500 km²—for each side)
2. the lethal area of the weapon on each missile (which I calculated to be 3.5 km²).

Thus, the fraction surviving, P_s, for side A can be taken as

$$P_s\left(A_{dep}\right) = e^{-\left(\dfrac{B_{dep} \times 3.5}{2,500}\right)},$$

and for side B,

$$P_s\left(B_{dep}\right) = e^{-\left(\dfrac{A_{dep} \times 3.5}{2,500}\right)}.$$

So, to achieve its objective of 300 missiles surviving a first strike by side B,

$$A_{req} = \frac{300}{P_s\left(A_{dep}\right)}.$$

Similarly,

$$B_{req} = \frac{300}{P_s\left(B_{dep}\right)}.$$

The results of these calculations on a range of force sizes for both sides are summarized in Table 2.1. The "Requires" columns of the table show the number of missiles that the indicated side must deploy to ensure that 300 missiles survive, as a function of the other side's deployment.

The results show that, for the conditions posited previously (the deployment area is constrained to 2,500 km², and the lethal area of each attacking missile is 3.5 km²), no stable, symmetrical deployment is possible. That is, there is no deployment for which both countries are satisfied that 300 missiles will survive. Specifically, the number in the third column for each side is always larger than the number in the first column.

Enlarging the deployment area for both sides can help: Table 2.2 shows the results for side A when one assumes that the lethal area for each attacking missile is 3.5 km² and the deployment area is 3,000 km².

Given a total deployment area of 3,000 km², both A and B meet the requirement of 300 missiles surviving between deployments of more

Table 2.1
Total Missiles Required to Ensure that 300 Survive a First Strike, Smaller Deployment Area

For A to Ensure 300 Survivors			For B to Ensure 300 Survivors		
If B Deploys (no.)	Percentage of A Surviving[a]	A Requires (no.)	If A Deploys (no.)	Percentage of B Surviving[a]	B Requires (no.)
0	100.0	300	0	100.0	300
200	75.5	397	200	75.5	397
400	57.1	525	400	57.1	525
600	43.2	695	600	43.2	695
700	37.5	799	700	37.5	799
800	32.6	919	800	32.6	919

NOTE: The lethal area for each attacking missile is 3.5 km², and the deployment area is 2,500 km².

[a] $P_S(A_{dep})$ and $P_S(B_{dep})$, respectively, rendered here as percentages.

Table 2.2
Total Missiles Required to Ensure that 300
Survive a First Strike, Larger Deployment Area

If B Deploys (no.)	Percentage of A Surviving	A Requires (no.)
0	100.0	300
200	79.2	397
400	62.7	478
600	49.7	604
700	44.2	679
800	39.3	763
900	35.0	857
1,000	31.1	963
1,200	24.7	1,217

NOTE: The lethal area for each attacking missile is
3.5 km^2, and the deployment area is 3,000 km^2.

than 600 and less than 1,200 missiles. In other words, A can achieve its goal of 300 missiles surviving without having to deploy more missiles than B has deployed. Beyond 1,200 missiles deployed by B, the situation is unstable. Each side, to meet the requirement, must deploy more and more missiles.

This analysis showed that not all approaches to constraining strategic nuclear forces were desirable from the standpoint of survivability and stability. I did not claim that the concept of an agreement whereby each side deploys mobile missiles in a designated area would not work. But—and this is the central point—the permitted deployment area must not be too small. One might prefer a small area from the standpoint of verification. But it cannot be too small, lest the survivability of the retaliatory force be compromised. Neither side will tolerate a position of inferiority, and such an agreement would not be negotiable.

I had read and reread the book *The Strategy of Conflict* by Dr. Schelling and took to heart the most insightful statements, one of which was, "If your enemy is concerned about the vulnerability of his forces, you should also be concerned."[4] This made me think about alternative ways to constrain nuclear forces. What if the agreement were to dictate that each side was constrained to deploy no more than a stated amount of destructive power (as defined by the quantity $ny^{2/3}$, where n represents the number of weapons deployed and y the yield of each weapon). One way to limit (deliverable) destructive power is to limit the throw-weight of one's missiles; throw-weight is a function of the volume of the missile. Only so much destructive power can be delivered by a missile of a given size (volume). How well you do is up to the engineers—rocket designers, as well as weapon designers.[5]

Now, suppose that one side (or both) chooses to deploy large missiles that have eight units (or erdels) of destructive power per missile. If each side is allowed no more than 800 erdels total, then each could deploy a total of 100 missiles.[6] On the other hand, one or both sides could choose to deploy smaller missiles that each delivered one erdel of destructive power. In this case, each side could deploy 800 missiles. I contended that, in a world in which each side's force is based at fixed sites, each side would choose the smaller missile. The aggregate destructive power is the same with either deployment (800 erdels), but the survival potential is greatly increased in the case of the smaller missiles— simply because the 800 erdels of kill potential are employed against 100 aimpoints (in the case of large missiles) and against 800 aimpoints (in the case of the smaller missile). The total destructive power is the same, but the kill potential per aimpoint is reduced by a factor of eight in the case of the smaller missile.

When I presented my analysis of the proposal by the two professors along with my own suggested approach, several of those at

---

[4] Thomas C. Schelling, *The Strategy of Conflict,* Cambridge, Mass.: Harvard University Press, 1960.

[5] See Glenn A. Kent, "On the Interaction of Opposing Forces Under Possible Arms Control Agreements," Cambridge, Mass.: Harvard University Center for International Affairs, Occasional Paper No. 5, March 1963.

[6] I assumed that the number of warheads per missile was always one. Multiple independently targetable reentry vehicle (MIRV)–outfitted missiles had yet to be invented.

the meeting objected, pointing out that the yield of the warhead on a Soviet missile cannot be verified without using extremely intrusive measures, which were unlikely to be negotiable. My reply was that it would not be necessary to verify the actual yield of Soviet weapons. Rather, I said, we were simply trying to constrain the aggregate (total) deliverable destructive potential, and for this purpose we could assume that the missiles of both sides had roughly the same ratio of destructive power to volume. If we are content to assume that both sides' rockets are roughly equal in efficiency, the verification task devolves to simply verifying the volume of the missile. In answer to the question, "How do you 'verify' volume?" I replied that we can use national technical means to determine the basic dimensions of the missiles and then apply a well-tested equation for the volume of a cylinder: $V = \pi r^2 h$, or volume equals 3.14 times the radius squared times the height.

The logic of my case was lost on many members of the seminar. They seemed to come to the totally illogical conclusion that since volume does not *exactly* define destructive power, there was no purpose in constraining it. This position flew in the face of the fact that, given the same level of technology, there was likely to be six times the destructive power in a missile with six times the throw-weight and that most other proposals on the table did not constrain missile size, only missile numbers.

The idea of constraining the total volume that each side could deploy gained some traction then but not enough to change what became the principal metric of arms control in the Strategic Arms Limitation Treaty (SALT) talks, namely, the number of missiles (irrespective of size) that each side could deploy.[7] This led to the situation in which a huge Soviet SS-18 counted no more than a much smaller U.S. Minuteman, even though the destructive power of the two is very, very different.

---

[7]   Later, after MIRV-outfitted missiles were invented, some analysts went to great lengths to prove the obvious: The volume of a missile does not necessarily define its destructive power $(ny^{2/3}/c^2)$. For a stated number of warheads (n), smart engineers can provide more yield per warhead and a better circular error probable (CEP).

Those who had the burden of "verifying" any agreement wanted an agreement that was "verifiable." Thus, and understandably, they focused on the number of missiles, which, being silo-based, could be determined simply by counting the silos. They regarded the requirement to verify the volume of each type of missile as an added (and unnecessary) complication. Again, they seemed to ignore the statement that they were not obliged to verify the actual destructive power of each system. Under my proposal they would have to verify only each missile's volume.

From SALT to START

Now we fast-forward to the early 1970s. The National Security Advisor (Henry Kissinger) directed DoD to evaluate four alternative arms control agreements in preparation for determining the U.S. position in the upcoming SALT talks. The four options all focused on constraining the number of ICBMs that each side could deploy.

Paul Nitze, the former Deputy Secretary of Defense, was serving as Secretary of Defense Melvin Laird's representative for SALT. Mr. Nitze directed me to head an interagency group to evaluate these options. The group consisted of

- two representatives from each of the services
- two from the JCS
- one from the Office of Program Analysis and Evaluation (PA&E)
- two from ACDA
- two from the U.S. Department of State
- two from DIA
- two from the Central Intelligence Agency.

So I was suddenly the head of a large and disparate group, each member with his own agenda.

There was quick agreement by all that option two was the chosen one. It fared better than all others according to various criteria that we had established. (One of those criteria was the ability to verify compliance with the agreement.) But there were several conditions. Predictably, the Navy wanted to make sure in our reply that we noted the inherent and increasing vulnerability of ICBMs in hardened missile silos, as well as the inherent survivability of SLBMs on stealthy submarines.

However, some of us were troubled by the list of options we were presented, noting that all of them (including option two) had the inherent problem of constraining the number of missiles while leaving the destructive power of each missile unconstrained. In a sense, option two was only the least bad of the four. Accordingly, I was inclined to report that none of the options were satisfactory.

I had little support from within the interagency group for this line of argument. The two representatives from the State Department, the two from ACDA, and the one from PA&E were the only members to agree. Later, after consultation with their superiors, the two from State and the two from ACDA withdrew their agreement. Nevertheless, I thought that some statement along the lines indicated should be made. To this end, I discussed this issue with Paul Nitze and made it clear that it was his call. He decided that it was important to point out the lack of satisfactory options and that a statement to this effect should remain as part of the evaluation.

So, without the support of most of the other members of the working group, I left in the report the statement I had drafted objecting to all the options. Mr. Nitze forwarded the package to the White House. The net effect, after some discussion with the staff of the National Security Council, was that the whole report went into "Deep Six." Obviously, the White House did not like the statement.

I had come close to having the issue debated at higher levels with a formidable advocate in Paul Nitze, but we were trumped when the whole report became null and void. I would have to wait for another day, another forum, and another champion.

That new forum and champion arrived in a rather unexpected way. In the early 1980s, there arose in Congress a group that I will call, for want of a better name, the "gang of six":

- Sen. Sam Nunn
- Sen. William Cohen
- Sen. Charles Percy
- Rep. Norman Dicks
- Rep. Albert Gore
- Rep. Les Aspin.

These men—four Democrats and two Republicans—gathered together for the purpose of promoting better arms control agreements. Their engagement in the issue was prompted, in part, by frustration with the Reagan administration's approach to arms control, which seemed (to them) to have been designed more with an eye toward irritating the Soviets than to actually gaining an agreement.

Senator Nunn had crafted a proposal called "builddown," the central tenet of which was to provide incentives for both sides to reduce their nuclear arsenals. His idea was to propose that whenever one side or the other chose to modernize an element of its strategic nuclear forces, the new delivery vehicles would replace the previous ones at the rate of one new to two old. That is, for every new missile or bomber deployed, two older ones would have to be scrapped.

At about the same time, the administration created the Scowcroft Commission. Headed by Brent Scowcroft, this group took on the task of defining the basing mode for the new M-X (Peacekeeper) missile.[8] They had crafted a paper that provided a rationale for deploying 50 M-X missiles in silos with ten warheads each. This was a daunting task, since the missiles would be vulnerable to a first strike; by concentrating more of the United States' nuclear capability in a small number of aimpoints, the deployment actually reduced first-strike stability.[9] Nevertheless, their paper was a masterpiece for the purpose intended,

---

[8]   The commission was necessary because, as a candidate for president, Ronald Reagan had declared his intention to cancel the planned deployment of the M-X in multiple shelters and on mobile launchers in the southwest United States.

[9]   The Scowcroft Commission issued its report in April 1983. For a critique of the report and trenchant observations about the role of arms control in U.S. defense strategy in the 1980s, see Thomas C. Schelling, "What Went Wrong with Arms Control?" *Foreign Affairs*, Vol. 64, No. 2, Winter 1985–1986, pp. 219–233.

which was centered on coming up with a rationale that was politically acceptable.

Senator Nunn, as head of the Senate Armed Services Committee, held hearings on this report. Gen David Jones (the former chairman of the JCS), Lt Gen Kelly Burke, and myself were invited to one of these hearings. By this time, I was working at RAND. We could be counted on to make favorable statements regarding the report by the Scowcroft Commission. At the hearings, Senator Nunn took the opportunity to ask each of us for our reaction to his builddown proposal. I was the third in line. General Jones made some favorable remarks, as did General Burke. Now it was my turn. I did not want to use this forum to provide a critical evaluation of the Nunn proposal, but neither did I want to give an unequivocal endorsement to the idea. I was saved by the bell: The senator's time expired before I had to speak.

The next day, I received a call from the senator. "I noticed yesterday," he said, "that when given the opportunity to endorse my proposal, you were silent. Why?" My reply was that I had serious doubts on two counts:

- There was no forcing function. If one side or another replaced an existing missile with a new missile, the trade was two to one: one new missile for two of the old. But there was nothing in the proposal that would ensure a builddown of missiles. In fact, the agreement would create incentives not to modernize one's forces.
- Like SALT, the proposal focused on the number of missiles and not on their destructive capacity. It would allow each side to modernize with new and larger missiles—as long as they scrapped two older ones. This could lead to less first-strike stability in the years hence.

At this point, while we were still on the phone, the senator asked me to come to his office that afternoon to go over these points. I arrived to be greeted by the senator along with several members of his staff and staffers of the other members of the gang of six. I thought I was wading into deep water, but such was not the case. After a discussion of less than an hour, the senator stated, "I am convinced you are right on both counts. You have told me what is wrong with my proposal. Now

I'd like to ask you to craft a new proposal that reacts to these points." "I would be delighted," I told him.

I went back to RAND with a new sense of purpose. I had the opportunity to provide inputs to some notable and influential champions—the gang of six. The first thing I did was to elicit the help of Randall DeValk and Edward Warner, my colleagues at RAND, and to inform the President of RAND (Dr. Donald Rice) of the turn of events. Randy, Ted, and I went to work and we came up with a new approach to arms control. The approach we advocated centered on placing a ceiling on each side's weapon stations. Each strategic nuclear delivery vehicle would be charged with a certain number of weapon stations, based on the volume of the missile divided by some constant to be determined or on the number of separate RVs that had been tested on that missile (whichever was larger).[10] As we refined our framework, our work was abetted by Arnold Punaro, who worked for Senator Nunn, and by the senator himself. The senator made one condition—the proposal would be known as the "builddown approach." We agreed to this, of course, and even improved on it by calling the new proposal "double build-down," in recognition of the fact that it would lead to reductions not only in launchers but also in overall destructive capacity.

Somewhat out of the blue, Ted and I were asked to present our approach to some people who were to gather at the home of James Woolsey. Woolsey, who had served as Under Secretary of the Navy, was also a member of the Scowcroft Commission. We were somewhat taken aback when we arrived and found that Scowcroft himself, as well as Congressman Aspin and other notables, were in attendance. In time, Senator Nunn's strategy for advancing our proposal became clear. First, he intended to oversee development of a proposal that all the gang of six would endorse and support. Next, he would take it to the Scowcroft Commission and gain their endorsement. Once this had been accomplished, he would go to the President and seek his approval.[11]

---

[10]   See Glenn A. Kent, Randall J. DeValk, and Edward L. Warner III, *A New Approach to Arms Control*, Santa Monica, Calif.: RAND Corporation, R-3140/FF/RC, 1984.

[11]   For a more detailed account, see Strobe Talbott, *Deadly Gambits: The Reagan Administration and the Stalemate in Nuclear Arms Control*, New York: Knopf, 1984.

Tables 2.3 and 2.4 show, respectively, U.S. and Soviet strategic nuclear forces as they were deployed in 1983, along with our determination of the number of standard weapon stations that should be associated with the forces of each side. The tables show that the forces of both sides, in aggregate, were roughly comparable in their destructive potential, with the U.S. ballistic missile force being smaller than the Soviets' but with the difference being made up largely by a bigger U.S. bomber force. (Not counting retired B-52 bombers, the destructive potential of the U.S. nuclear arsenal was 13,656 standard weapon stations; that of the Soviet Union was 15,986.) Our proposal was that both sides agree to reduce their arsenals to a lower, common ceiling denominated in standard weapon stations.

It all worked. Finally, President Ronald Reagan gave his endorsement to the "double builddown" approach at a ceremony in the Rose Garden. The head of the U.S. delegation to the Strategic Arms Reduction Treaty (START) talks was instructed to describe the proposal to the Soviets at the next meeting in Geneva. James Woolsey was to go along to help. Unfortunately, for reasons unrelated to the merits of double builddown, the Soviet delegation walked out of these meetings. They had been instructed by Moscow to do so prior to the session as a demonstration of Moscow's pique over the deployment of new U.S. intermediate nuclear forces to Europe.

Within a few years, there was a distinct thaw in the Cold War. By the late 1980s, arms control agreements with the Soviet Union were suddenly quite feasible. By then, however, I had turned to other matters. To my knowledge, the concept of counting weapon stations (and other features of the "new approach" outlined in Kent with DeValk and Warner, 1984) were never incorporated into any formal agreements. Oh well; be that as it may, I gave it a try for nigh on 25 years or so.

What should we learn from this? First, that one should strive to define clearly the objectives of specific policy initiatives. The decades-long pursuit of agreements to control the deployment of strategic nuclear weapons with the Soviet Union was plagued, in part, by the fact that different bureaucratic entities had different policy objectives in mind when they approached the topic of arms control. Some sought to contribute momentum in the overall process of détente; others were

Table 2.3
U.S. Strategic Forces, Mid-1983

Force	Number	Missile Throw-Weight (000s kg)		Actual or Estimated Weapons	SALT Weapons		Standard Weapon Stations	
		Per Missile	Total		Per Missile	Total	Per Missile	Total
Ballistic missiles								
ICBMs								
Titan II	45	3.8	171		1	45	7.6	342
Minuteman II	450	0.7	315		1	450	1.4	630
Minuteman III	550	1.0	550		3	1,650	3.0[a]	1,650
Total	1,045		1,036	2,100		2,145		2,622
SLBMs[a]								
C-3	304	1.5	456		14	4,256	14.0[b]	4,256
C-4	264	1.3	343		8	2,112	8.0[b]	2,112
Total	568		799	5,200		6,368		6,368
Total ballistic missiles	1,613		1,835	7,300		8,513		8,990

Table 2.3—Continued

Force	Number	Bomber Takeoff Gross Weight (000s lbs)		Actual or Estimated Weapons	SALT Weapons		Standard Weapon Stations	
		Per Bomber	Total		Per Bomber	Total	Per Bomber	Total
Bombers (active)								
B-52G with ALCMs	104	488	49,752				19.5	2,028
B-52G	66	488	32,208				9.8	647
B-52H with ALCMs	95	488	46,400				19.5	1,853
FB-111A	60	115	6,900				2.3	138
Total active bombers	325		135,260	2,900				4,666
Retired B-52s[d]	308	488	150,300				9.8	3,018
Total bombers	633		285,560					

Force Summary	Total Number	Actual or Estimated Weapons	Standard Weapon Stations
Missiles and active bombers	1,938	10,200	13,656
Adding in retired bombers	2,246	10,200	16,674

SOURCE: Adapted from Kent, DeValk, and Warner, 1984, pp. 30–31.

NOTE: The data in this table reflect the status of forces at the time the source document was published.

a The U.S. submarine force consisted of 31 Poseidon submarines, of which 19 were equipped with C-3s and 12 with C-4s, and three Trident submarines equipped with C-4s.

b The number of standard weapon stations declared or tested exceeded the value obtained by applying the missile throw-weight or bomber takeoff gross weight counting rules.

c These are SALT-accountable.

Table 2.4
Soviet Strategic Forces, Mid-1983

Force	Number	Missile Throw-Weight (000s kg)		Actual or Estimated Weapons	SALT Weapons		Standard Weapon Stations	
		Per Missile	Total		Per Missile	Total	Per Missile	Total
ICBMs								
SS-11	550	0.9	495		1	550	1.8	990
SS-13	60	0.5	30		1	60	1.0	60
SS-17	150	2.7	405		4	600	6.7	1,005
SS-18	308	8.0	2,464		10	3,080	20.0	6,160
SS-19	330	3.6	1,188		6	1,980	9.0	2,970
Total	1,398		4,582	5,700		6,270		11,185
SLBMs[a]								
SS-N-6	384	0.7	269		1	384	1.4	538
SS-N-8	292	0.7	204		1	292	1.4	409
SS-N-18	224	1.0	224		7	1,568	7.0[b]	1,568
Total	900		697	1,800		2,244		2,515
Total ballistic missiles	2,298		5,279	7,500		8,514		13,700

Table 2.4—Continued

Force	Number	Bomber Takeoff Gross Weight (000s lbs)		Actual or Estimated Weapons	SALT Weapons		Standard Weapon Stations	
		Per Bomber	Total		Per Bomber	Total	Per Bomber	Total
Bombers								
Tu-95 Bear	100	414	41,400				8.3	830
Mya-4 Bison	43	350	15,050				7.0	301
Tu-22M Backfire	210	277	58,170				5.5	1,155
Total	353		114,620	1,200				2,286

Force Summary	Total Number	Actual or Estimated Weapons	Standard Weapon Stations
Missiles and active bombers	2,651	8,700	15,986
Adding in retired bombers	—	—	—

SOURCE: Kent, DeValk, and Warner, 1984, pp. 32–33.

NOTE: The data in this table reflect the status of forces at the time the source document was published.

[a] The Soviet submarine force consisted of 14 Delta III submarines, each equipped with 16 SS-N-18s; four Delta IIs, each with 16 SS-N-8s; 18 Delta Is, each with 12 SS-N-8s; one Hotel III with six SS-N-8s; one Golf III with six SS-N-8s; and 24 Yankee I submarines, each with 16 SS-N-6s.

[b] The number of standard weapon stations declared or tested exceeds the value obtained by applying the missile throw-weight or bomber takeoff gross weight counting rules.

interested in slowing the momentum of the "arms race" as a means of channeling the U.S.–Soviet competition into other, perhaps less dangerous fields or simply to save money. I always felt that reducing the chances of nuclear war should be our primary objective, but oddly, this view was not shared by many others. Second, one should strive to get one's analysis and its implications injected into the policy debate early, before minds are made up and positions are set. It is easier to influence policy when things are inchoate than it is after specific positions have been publicly articulated.

In spite of all our work, the idea of paying attention to the vulnerability of the force of each side to a first strike never was formally enshrined as a critical element of many agreements with the Soviets, although the notion that arms control agreements could and should be designed to strengthen survivability was. Hence, first-strike stability did gain traction in some quarters.

"Stability" Between U.S. and Soviet Strategic Forces

President Reagan's call to develop defenses that would render nuclear weapons "impotent and obsolete" fueled a debate that had been dormant since the 1972 ABM treaty.[12] That treaty codified the reality that I had illuminated in the 1964 damage-limiting study, namely, that the U.S.–Soviet strategic relationship would remain dominated by offensive nuclear forces and that heavy investment in strategic defenses was

---

[12] Ronald Reagan, "Address to the Nation on Defense and National Security," March 23, 1983. I published several studies at RAND on stability and the U.S.–Soviet strategic relationship. See Glenn A. Kent and Randall J. DeValk, *Strategic Defenses and the Transition to Assured Survival*, Santa Monica, Calif.: RAND Corporation, R-3369-AF, 1986; Glenn A. Kent, "A Suggested Policy Framework for Strategic Defenses," Santa Monica, Calif.: RAND Corporation, N-2432-FF/RC, 1986; Glenn A. Kent, Randall J. DeValk, and David E. Thaler, "A Calculus of First-Strike Stability: A Criterion for Evaluating Strategic Forces," Santa Monica, Calif.: RAND Corporation, N-2526-AF, 1988; Glenn A. Kent and David E. Thaler, *First-Strike Stability: A Methodology for Evaluating Strategic Forces*, Santa Monica, Calif.: RAND Corporation, R-3765-AF, 1989; and Glenn A. Kent and David E. Thaler, *First-Strike Stability and Strategic Defenses: Part II of a Methodology for Evaluating Strategic Forces*, Santa Monica, Calif.: RAND Corporation, R-3918-AF, 1990.

both financially and strategically counterproductive. The ABM treaty therefore limited BMD to very small numbers and laid the groundwork for arms control talks aimed at capping or reducing the number of U.S. and Soviet nuclear weapons.

Reagan's televised speech on March 23, 1983, unveiled a Strategic Defense Initiative (SDI) with the aim of moving the U.S.–Soviet relationship toward a state in which strategic defenses dominated strategic offenses, at least as far as ballistic missiles were concerned. At the time of the speech, the United States and the Soviet Union were engaged in arms control negotiations that would lead to START I in 1991. The SDI complicated these talks.

In large part, the speech touched off a debate over the impact of strategic defenses on the "stability" of the U.S.–Soviet nuclear relationship. Advocates of strategic defenses criticized U.S. defense policy for relying on deterrence—and the underlying threat of catastrophic damage to the Soviet Union that underpinned it—as the primary means of keeping the United States safe from nuclear attack. Instead, they argued, it would be better to be able to defend against a Soviet attack, relying on U.S. capabilities rather than on Soviet perceptions and self-restraint.[13] They contended that strategic defenses would help ensure strategic stability between the United States and the Soviet Union.[14] These arguments were superficially appealing. But having thought extensively about what I called *first-strike stability*, I believed that such blanket statements were wrong-minded and set out to shed light on the issue of stability through rigorous analysis. Later in this chapter, I return to the issue of stability and strategic defenses; first, I want to expand on the concept of first-strike stability itself.

The Concept of First-Strike Stability

During the Cold War, deterrence was the calculus against which strategic nuclear forces traditionally were evaluated. Deterrence depended

---

[13] See, for example, Daniel O. Graham and Gregory A. Fossedal, "A Defense That Defends," *Wall Street Journal*, April 8, 1983.

[14] Fred Charles Iklé, "Nuclear Strategy: Can There Be a Happy Ending?" *Foreign Affairs*, Vol. 63, No. 4, Spring 1985, pp. 810–826.

on the U.S. capability to cause "unacceptable damage" to the Soviet Union in response to an attack on the United States by Soviet strategic nuclear forces. According to this construct, the United States sought to ensure that the leaders of the Soviet Union would conclude that the "cost" of a U.S.–Soviet strategic nuclear exchange to the Soviet Union (in terms of damage to its "value structure") would always exceed any benefit (in terms of damage to the U.S. value structure) that the Soviets might gain from initiating such an exchange. In this context, the Soviets' value structure included the population, war-supporting industries, theater projection forces, and its national leadership. Because the United States could hold these assets at risk with a fairly modest number of weapons, it was easy to make the case that the cost to the Soviet Union of initiating a first strike would be far greater than zero—zero being the cost of not initiating a strike in the first place, or the status quo.

I always had a nagging concern about this construction in which the Soviet cost of going first was compared to the Soviet cost of the status quo. Not only was the first-strike versus status quo comparison troubling, it did not consider the possibility of a first strike *by the United States* in the minds of Soviet leaders; thus, deterrence was, in effect, a one-sided calculus. I believed that the calculus must be two-sided. In a crisis, the leaders of the Soviet Union might compare the cost of striking first with the cost of waiting and going second *after absorbing a U.S. first strike*, and that mirror construct would apply equally to U.S. leaders. In this construct, the expected cost of waiting was not necessarily zero. Rather, it was the cost of going second multiplied by the probability that such an event would come about. Were there a great difference between the cost to both sides of going first and going second, there would, in a crisis, be a self-feeding, unstable situation. Each side would have incentive to go first because of the belief that, in waiting, it was apt to incur a first strike by the other side.

This concept was not all that new. Thomas Schelling first elucidated it in the early 1960s in his seminal book, *The Strategy of Conflict*, in which he wrote that "we live in an era in which a potent incentive on either side . . . to initiate total war . . . is the fear of being a poor second for not going first."[15] Schelling focused, in particular, on the posture of strategic nuclear forces as a factor in a decision to strike first or wait and emphasized the two-sided nature of the strategic relationship:

> [The enemy's] manifest invulnerability to our first strike could be to our advantage if it relieved him of a principal concern that might motivate him to try striking first. If *he* has to worry about the exposure of *his* strategic forces to a surprise attack by *us*, *we* have to worry about it too.[16]

I made a clear distinction between *crisis instability* and *first-strike instability*. Crisis instability would arise from numerous factors that might induce instability in a crisis, such as psychological stress, ambiguous or incorrect information, and erroneous assessment of enemy intent. The concept of first-strike stability focused only on the structure and posture of the strategic nuclear forces of both sides. As such, it was one element of crisis stability. My reasoning was that, of the many factors that might operate to cause a crisis to get out of hand, we (and the Soviets) should ensure that the posture of the strategic nuclear forces *of both sides* would not be prominent among them. Thus, first-strike instability was a potential cause of crisis instability, which was a much broader issue. Importantly, I did not believe that the latter was quantifiable; I set out (with the aid of Randy DeValk and David Thaler, two research assistants at RAND) to quantify the former (first-strike stability).[17]

Our main focus was on comparing structures and postures of strategic nuclear forces according to the weapons remaining to a first-striker after attacking the victim's offensive forces in a maximum effort

---

[15] Schelling, 1960, p. 231.

[16] Schelling, 1960, p. 238; emphasis in the original.

[17] At the same time, two other teams at RAND were working on various aspects of stability. Dean Wilkening and Ken Watman published a study, *Strategic Defenses and First-Strike Stability*, Santa Monica, Calif.: RAND Corporation, R-3412-AF, 1986. Russ Shaver and Jim Thomson conducted a study of their own. We had many very useful interactions with these other teams, and these improved our analysis immensely.

to limit damage to its own value.[18] I employed "drawdown" curves, a tool I had developed as early as 1970, during my time as head of AFSA. These curves were a very transparent way of showing the "counterforce" options each side had in a first strike against the strategic nuclear forces of the other. I drew the curves in a graph to show Soviet weapons available on the x-axis and U.S. weapons available on the y-axis (a graph we termed the "weapons domain"). The first-striker would attack the most lucrative strategic offensive targets first and engage in other counterforce options in descending order of "return"—e.g., ballistic missile submarines (SSBNs) in port (the most lucrative), bombers on the ground, ICBMs in hardened silos, and bomber flyout areas or SSBN patrol areas (the least lucrative). In each option, the first-striker would destroy the other's weapons by expending its own, so the curves in the end defined how many of the victim's weapons were not destroyed (and were available to retaliate against the first-striker's value structure) and how many of the first-striker's weapons were not expended in counterforce (and were available against the victim's value structure). With the least lucrative options, the "return" in terms of limiting damage to the first-striker's value was so small that the first-striker might find it best to forgo these options and, instead, hold these weapons in reserve or use them to attack the victim's value structure.[19]

I used these curves in this initial effort to gain a sense of the potential stability inherent in the structure and posture of alternative U.S. and Soviet strategic nuclear forces. Figure 2.3 shows notional drawdown curves under three corresponding hypothetical situations: (1) one in which both sides have highly survivable nuclear forces, and many retaliatory weapons would be available to each even after absorbing the other's first strike; (2) one in which the Soviets can greatly reduce U.S. retaliatory capability and possibly limit damage to Soviet value to a significant degree; and (3) one in which *both* sides can limit

---

[18]   See Kent, DeValk, and Thaler, 1988.

[19]   See the explanation of drawdown curves in Kent, DeValk, and Thaler, 1988, pp. 6–15. We used "standard weapon stations" in this study in lieu of actual weapons because of my arms control work with Ted Warner. See "Changing the Paradigm of Arms Control," pp. 56–72, for an explication of this metric. In later reports on first-strike stability, I portrayed actual numbers of weapons.

Figure 2.3
Drawdown Curves in the Weapons Domain with Areas
of First-Strike Instability

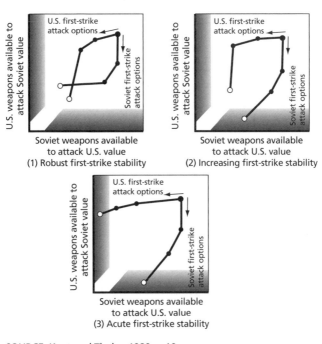

SOURCE: Kent and Thaler, 1989, p.10.
RAND OP223-2.3

damage to themselves considerably by striking first. Note that we portrayed shaded "keep-out zones" along the two axes. These were notional areas to be avoided to maintain "robust" first-strike stability. In other words, if both sides could, in a first strike, reduce the other's retaliatory capability to points within the keep-out zones, a condition of "acute" first-strike instability would presumably exist.

While this initial effort helped us depict the potential effects of changes in strategic offensive force posture and structure, the direct effect on first-strike stability remained quite notional. We decided to take the next step of quantifying these keep-out zones and developing an "index" of first-strike stability that would enable clear comparisons between alternative structures and postures.[20]

My first task was to define *cost*. After some deliberation, I developed an equation for each side that incorporated the damage that side incurs in a nuclear exchange—in terms of the percentage of value destroyed—plus the damage not inflicted on the other side. Since each side would place more emphasis on limiting damage to its own value than inflicting damage on the value of the other side, we discounted the second factor by an appropriate coefficient. The form of cost for each side, therefore, was

$$C_{sideA} = D_{sideA} + \lambda\left(1 - D_{sideB}\right),$$

where C is cost, D is damage (in percentage of value), and λ is the coefficient, which I set at 0.3 (although the methodology could accommodate any preference). Thus, if the damage each side incurred to its value in a nuclear exchange was 70 percent, the cost to each side would be 0.70 + 0.30 (1 − 0.70) = 0.70 + 0.09 = 0.79. I also noted that, by the above calculus, the cost of a war was always positive, i.e., there was no "benefit" (negative cost) to war, and the best either country could possibly achieve was a "cost" of zero. This happened when one side suffered no damage and inflicted 100 percent damage—a quite unlikely case. Thus, according to my construct, the cost of a nuclear exchange would be greater than the cost of neither side using nuclear weapons.

I calculated the percentage of the value structure damaged with the help of a plot I constructed for each side, establishing the relationship between weapons available to be employed against value (that is, those weapons not used to attack strategic offensive forces in an effort to limit damage to oneself) and value damaged.[21] As Figure 2.4 shows, the curve for each side was concave downward to reflect the point that, as more and more weapons are employed against the value of the other side, they are used on less-lucrative targets and thus for diminishing

---

[20] See Kent and Thaler, 1989.

[21] See Kent and Thaler, 1989, p. 18, Figure 7.

Figure 2.4
Relationship Between Weapons Delivered and Value Damaged,
United States and Soviet Union

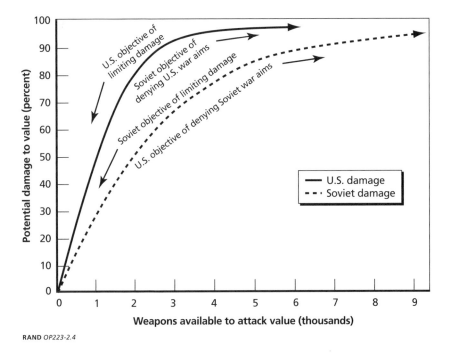

returns. We noted also that U.S. value was more concentrated than Soviet value.[22] So, if one knows the weapons available to attack value, one can determine the damage to value and thus cost. In fact, it was now possible to plot lines of "constant cost" in the weapons domain and to determine the cost to each side of striking first or second using the drawdown curves for each structure and posture of U.S. and Soviet strategic forces.[23]

To be able to compare the relative first-strike stability (or instability) in these structures and postures, we took the final step of combining the four costs—i.e., for each pair of U.S. and Soviet force structures

---

[22] So, 2,000 weapons could damage 80 percent of U.S. value and only about 55 percent of Soviet value. See Kent and Thaler, 1989, pp. 17–19.

[23] See Kent and Thaler, 1989, pp. 20–24.

and postures, the costs to each side of going first and second—into a single index. My first-strike stability index was the product of

$$\frac{C_1^{U.S.}}{C_2^{U.S.}} \times \frac{C_1^{Sov}}{C_2^{Sov}}.$$

The index was thus bounded by zero and one. If either side could reduce the cost of the exchange to zero, the index would have a value of zero—a completely unstable situation. On the other hand, if neither side could muster any "damage-limiting" capability at all, the cost to each side of striking first would equal its cost of waiting and retaliating; each ratio would be one, and the product would be one—first-strike stability would be as good as it could be. Generally, however, the index would be greater than zero and less than one. Normally, we used the index to compare alternative forces, not to gain any truths about the absolute level of stability for a particular case. This index would apply equally to the United States and the Soviet Union; i.e., there was no U.S. index separate and distinct from a Soviet one.

Insights on Strategic Offensive Forces

Armed with this index, we were able to gain considerable insight into the interaction of strategic nuclear forces. The index provided a means by which we could assess postures of existing forces, as well as the effects of various arms control proposals. Some of these findings were rather surprising.

First, we were able to show that, contrary to some arguments being made at the time, the U.S. and Soviet strategic forces were rather robust in terms of first-strike stability. The structures and postures of the two sides were such that neither side was likely to perceive much advantage in striking first. Each side had enough survivable retaliatory capability (even at day-to-day alert levels) to prevent the adversary from significantly limiting damage to itself through first-strike counterforce options. Moreover, we demonstrated that generating forces to make them more survivable—such as putting more SSBNs to sea, dispatching mobile missiles out of garrison, or putting more bombers on strip alert—would improve first-strike stability. The popular conception was (and remains to this day) that placing forces on alert makes it more likely that war will break out. But generating forces makes them less vulnerable to the adversary's first strike, which decreases one's cost of waiting while increasing the enemy's cost of going first.

This led us to make a seemingly counterintuitive recommendation that U.S. and Soviet leaders begin to view the generation of forces early in a crisis as *stabilizing* because it helps further remove incentives for either side to strike first. At the same time, we emphasized that forces already survivable (or generated) on a day-to-day basis were preferable to those requiring action by leaders in a crisis to guarantee their survivability. This is so because no decision or special actions were needed to make such forces more survivable.

Second, we demonstrated that expanding inventories of strategic nuclear forces is not necessarily *destabilizing* and that reducing inventories through arms control is not necessarily *stabilizing*. These conclusions were of course quite disquieting to the arms control community. But we were correct: As long as each side maintained a large retaliatory capability on a day-to-day basis in the form of survivable, nontargetable weapons, the fact that inventories (and weapons available for counterforce) are increased would be irrelevant. (However, we did acknowledge that an arms race would negatively affect other elements of the U.S.–Soviet relationship.) On the other hand, I found that proposals for arms reductions under START could actually increase first-strike instability unless steps were taken to increase day-to-day force generation rates. For example, reducing the number of SSBNs, without increasing the percentage kept at sea day to day, would undercut a very important part of each side's retaliatory capability; banning land-based mobile missiles would have a similarly deleterious effect.

There was a rather shocking finding that we decided not to emphasize: Some forms of cheating on arms control agreements could actually be first-strike stabilizing! Put simply, if each side knew the other had additional, treaty noncompliant forces but did not know where they were, those forces would by definition be nontargetable and survivable.

Finally, we placed first-strike stability in the broader context of U.S. national security objectives and recommended that strategic nuclear forces be evaluated in part according to their effects on first-strike stability. This was particularly important because first-strike stability was in some conflict with other important objectives, particularly limiting damage and providing "extended deterrence" to our allies.

As I have explained earlier, the core of first-strike stability was the inability of either side to limit damage to any significant degree to itself in a first strike, so pursuit of damage-limiting options necessarily would lower the index. As for extended deterrence, this was based on encouraging a Soviet belief that actions they might take that were severely detrimental to U.S. interests (such as invading our NATO allies) would present a grave danger of unwanted and uncontrollable escalation; yet first-strike stability relieved pressure on both sides to strike first in a deep crisis. So, in modernizing forces and formulating arms control proposals, U.S. policymakers should consider trade-offs between these competing national security objectives.

The Debate over Strategic Defenses

In his famous 1983 speech about SDI, President Reagan asked,

> What if free people could live secure in the knowledge that their security did not rest upon the threat of instant U.S. retaliation to deter a Soviet attack, that we could intercept and destroy Soviet strategic ballistic missiles before they reached our own soil and that of our allies?[24]

The President's question underscored the point that, although the Soviet Union had been deterred from attacking us for some 35 years, our national survival had, in the final analysis, depended on the restraint and calculus of gains and risks of our Soviet adversary, especially in a crisis. We did not fully control our own destiny. In short, the President had set a new goal of attaining a condition of U.S. "assured

---

[24] Reagan, 1983.

survival" through the development and deployment of nationwide BMD.[25] To say the least, this was a lofty and daunting goal.

Subsequently, Randy DeValk and I undertook an effort to provide some insight into the potential perils along the way as the United States, and perhaps the Soviet Union as well, proceeded from the extant posture of no nationwide defenses to one in which one or both deployed highly effective nationwide defenses. This was a journey that was liable to take many years and have many twists and turns.

The Transition from Assured Destruction to Assured Survival

When we started the analysis, the "concept" of these defenses had yet to be defined in any detail. The presumption by many was that they would be space-based (hence the name "Star Wars"). There would be "battle stations" (satellites in low earth orbit [LEO]) that contained sensors and interceptors with "hit-to-kill" capability. My purpose was neither to attempt to define the concept in any detail nor to evaluate its effectiveness. Rather, my effort was directed at more-strategic questions: What were the strategic implications if one or both countries could actually deploy highly effective defenses?

I defined the defenses in terms of their "defense potential"—the number of ballistic missile RVs, or warheads, that the defense could "subtract" from an attack by the other side. One's defense potential could operate to blunt or negate an adversary's counterforce attack (in the adversary's first strike) or an adversary's retaliatory attack after absorbing a first strike. Though a rather gross measure, this probably was adequate for the purpose intended and in the absence of an official definition of the concept of the defense. In addition, we differentiated *assured* survival from *conditional* survival. Assured survival was the situation in which a nation had a defense potential equal to or exceeding the inventory of ballistic missile RVs deployed by the adversary. Alternatively, conditional survival was a situation in which a nation's defense potential would equal or exceed the number of RVs in a retal-

---

[25] See Kent and DeValk, 1986. I assumed that the Soviet Union would also embark upon this journey—a most dubious assumption given the hindsight available to those in the post–Cold War world.

iatory attack—but not a first strike—by the adversary. In this case, the nation's survival was attained only on condition of a first strike. As such, conditional survival could lead to first-strike instability and create incentives for leaders to strike first in a deep crisis.

Our calculations mirrored those we had used to build drawdown curves, except that the number of ballistic missile RVs "arriving" at their intended targets was reduced by the BMD of the other side. Table 2.5 gives a notional posture and capability of the strategic offensive forces of both sides. It shows (1) the total number of RVs deployed by each side, including those atop both ICBMs in silos and SLBMs aboard invulnerable SSBNs at sea, (2) the number and effectiveness of "killer" RVs that are useful in a counterforce attack against the other's ICBM-borne RVs in hardened silos, and (3) the number of ICBM silos that each side possesses. Note that the Soviets were assumed to have a much better counterforce capability—5,000 Soviet killer RVs, each with a P_k of 0.7 against 1,000 U.S. ICBM silos—meaning that the Soviets could, in the absence of defenses by the United States, destroy 70 percent of U.S. silos with 1,000 RVs. This capability was in comparison to 1,500 U.S. killer RVs with a P_k of 0.4 against 1,400 Soviet silos (see Table 2.5).

Given these postures and capabilities, Table 2.6 presents the calculus for U.S. and Soviet first-strike and retaliatory capabilities in the presence of a defense potential of 3,800 on each side. Table 2.6 shows that, in striking first, the United States could not limit damage to itself to any significant degree because it did not have enough killer RVs to penetrate Soviet defenses, and thus a Soviet retaliation could overwhelm U.S. defenses, with 3,200 Soviet RVs penetrating to attack U.S. value. On the other hand, if the Soviets struck first, they could destroy enough U.S. RVs to ensure that few would penetrate Soviet defenses in a retaliatory attack. Thus, this case—a defense potential of 3,800 on both sides—puts the Soviet Union in a zone of conditional survival in which it could "survive" only if it strikes first—at least as far as ballistic missiles are concerned. The United States, conversely, would not achieve conditional survival.

We developed a graphic called the "defense domain" that is depicted in Figure 2.5. The graphic shows U.S. defense potential on

Table 2.5
Notional U.S. and Soviet Deployments of Ballistic
Missiles

Deployments	United States	Soviet Union
Total ICBM RVs deployed	2,000	6,000
Killer RVs	1,500[a]	5,000[b]
SLBM RVs	3,000	1,000
Total RVs deployed	5,000	7,000
Number of silos	1,000	1,400

[a] $P_k = 0.4$.
[b] $P_k = 0.7$.

Table 2.6
Notional First-Strike and Retaliatory Capabilities

Capabilities	First Strike	
	United States	Soviet Union
Killer RVs in counterforce attack	1,500	5,000
Defense potential (of other side)	3,800	3,800
Killer RVs that penetrate	0	1,200
Adversary ICBM RVs that survive counterforce attack	6,000	800
SLBM RVs at sea (other side)	1,000	3,000
Total RVs in retaliatory attack	7,000	3,800
Total RVs that penetrate	3,200	0

NOTE: In the presence of a potential ballistic missile defense of 3,800.

the x-axis and Soviet defense potential on the y-axis. By repeating these calculations for various combinations of U.S. and Soviet defense potentials, we were able to create for each structure and posture of U.S. and Soviet strategic offensive forces a "map" of zones of assured survival and conditional survival. We could also show paths from the existing reality of mutual assured destruction to the ultimate goal of mutual

Figure 2.5
Zones of U.S. and Soviet Conditional Survival in a Defense Domain

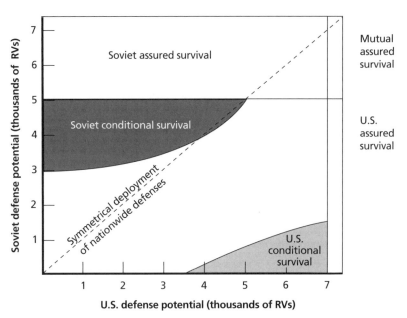

SOURCE: Kent and DeValk, 1986, p.15.
RAND OP223-2.5

assured survival that would avoid the danger zones of U.S. or Soviet conditional survival.

The work on the transition to assured survival led us to conclude that mutual assured destruction and mutual assured survival were the only two conditions that could offer both first-strike stability and arms race stability. We also concluded that it was possible to transition to mutual assured survival if both sides maintained their strategic defensive and offensive forces in highly survivable postures. The more survivable these postures, the smaller the U.S. and Soviet zones of conditional survival and the wider the path to mutual assured survival. On the other hand, increasing hard-target kill (HTK) capabilities while failing to improve survivability would tend to close this path.

We also noted that the comparatively large Soviet counterforce capability (as given in Table 2.6) meant that symmetrical deployment of intermediate levels of strategic BMD would erode the contribution of U.S. ballistic missiles to deterrence and thus could potentially decrease first-strike stability. To avoid such erosion during a transition, the United States would need to rely heavily on its strategic bombers to maintain an adequate retaliatory capability. Moreover, given the strategic offensive forces both sides possessed in the 1980s, the United States would have needed to build and deploy strategic BMD potential at nearly twice the rate of the Soviet Union to ensure that the larger zone of Soviet conditional survival could be avoided.

Calculating First-Strike Stability in the Presence of Strategic Defenses

After developing the methodology for determining the relative first-strike stability of stated postures of strategic offensive and defensive forces, we decided to address the arguments of those who contended that deploying strategic defenses would improve stability. Our work on the transition to mutual assured survival provided the foundation. We could calculate the index of first-strike stability throughout the defense domain to show how building strategic defenses would affect the U.S.–Soviet strategic relationship. As before, we had to develop a separate defense domain for each combination of U.S. and Soviet offensive force structures and postures.

Figure 2.6 shows a plot in the defense domain with lines of constant value of the first-strike stability index for offensive forces circa 1989 in a moderately generated (i.e., between peacetime and fully generated) posture. The plot shows that first-strike stability was relatively robust under then-existing conditions of no strategic defenses on either side (U.S./Soviet defense potential = 0/0). It also shows that first-strike stability continued to be robust in the opposing corners of the domain—where (1) only one side deployed high levels of defense or (2) both sides did so. In the first case, this derived from the fact that the defenses of the side that deployed them could handle either the first strike or the retaliation of the side that lacked defenses, whereas the latter side would suffer severe damage (and high cost) whether it struck first or second. Thus, there would be little difference between striking first and waiting, rendering first-strike stability relatively robust. And in the case of mutual deployment of high levels of defense potential

Figure 2.6
Defense Domain with Isolines of Constant Values of the Stability Index

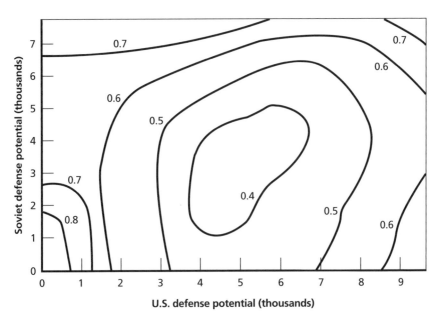

SOURCE: Kent and Thaler, 1989, p. 29.
RAND OP223-2.6

(the upper right corner of the domain), a condition of "defense dominance" would apply wherein the cost to each side would be low regardless of which strikes first. And with little difference in cost between striking first or second, the stability index is relatively high.

However, deploying "intermediate" levels of nationwide BMD—a defense potential of about 4,000 to 6,000 on the U.S. side and about 2,000 to 5,000 on the Soviet side—would create a "sinkhole" of relative first-strike instability, as shown in the center of the domain in Figure 2.6. At these levels, each side could substantially limit damage to itself by striking first, and thus, first-strike stability would erode. The so-called sinkhole tended to be deeper close to the axis representing U.S. defense potential because the Soviets had a small bomber force that exhibited low alert rates. Unlike the U.S. bomber force, the Soviet bomber force was inadequate to fill the void created by the low penetrability of Soviet RVs in a Soviet retaliatory strike.

Our conclusion was that advocates of deploying nationwide BMD *on the basis of enhancing first-strike stability* were decidedly wrong. We had little doubt that deploying defenses, especially in the context of U.S.–Soviet competition, would actually reduce first-strike stability and probably reduce incentives on either side to adhere to agreements that constrain or reduce offensive arms, including START I. This was certainly an argument that caught the attention of arms control proponents who had opposed SDI because of its deleterious effects on the atmosphere and substance of ongoing START talks between the United States and the Soviet Union.

On the other hand, it is doubtful that this conclusion swayed those in the administration who favored moving ahead with SDI. I remember briefing a senior official at the Strategic Defense Initiative Organization. I painstakingly went through the methodology we had developed and showed how defenses could actually reduce first-strike stability. I ended with the defense domain chart shown in Figure 2.6. The official then walked up to the front of the room, where the slide was projected, put his finger somewhere close to the middle of the chart (where first-strike stability was the lowest), and emphatically stated, "*This* is where we want to be." I was somewhat stunned, to say the least. It appeared to me at that point that the advocates of SDI were not to be swayed by any analysis that was not consistent with their agenda, the centerpiece of which was simply to deploy defenses, irrespective of the consequences for U.S. security.

Final Thoughts

My plea—that first-strike stability should be an avowed and explicit criterion and objective in evaluating not only the structure and posture of strategic nuclear forces but also in arms control negotiations—never gained the traction that I thought it deserved. In fact, sometimes quite the opposite occurred. Considerations of negotiability with the Soviets, verifiability, and salability to the Congress all confused the debate and made some senior officials reluctant to revisit decisions that had been made years before. Once locked into a concept of denominating a strategic arms control agreement in delivery vehicles (irrespective of their capacity), both the Nixon and the Carter administrations were reluctant to change. Certainly the Soviets, with their advantage in missile throw-weight, had no incentive to entertain proposals for a different approach. And by the time the Reagan administration came to power, the notion that we had an interest in maintaining first-strike stability became unfashionable to say the least.

For example, an interagency working group formed in 1985 or thereabouts to evaluate a new concept for arms control that I had developed along with Ted Warner decried our approach because it "allowed, and even encouraged" measures by the Soviets to make their strategic nuclear forces less vulnerable.[26] Only after vigorous protest did they delete this very wrong-minded statement from their critique. The idea that putting submarines to sea and placing bombers on quick alert should be viewed as "stabilizing" actions in a crisis was adopted "in principle" by some, including the commander of STRATCOM. But, to my knowledge, this was never really implemented.

The end of the Cold War and the resulting shift in U.S. defense priorities placed our considerations of first-strike stability in a new context. Without a peer competitor, U.S. defense planners are understandably less concerned (even unconcerned) about first-strike stability. Still, the work I did on quantifying and gaining insight into this once-central objective does have implications for today's environment. It provides a way of thinking about stability between actors with smaller arsenals— for example, India and Pakistan. It is also useful for future situations in which multiple actors with nuclear weapons face one another.[27]

Obtaining one value for the index required quite a lot of calculating to determine counterforce options, damage to value, and cost. It is important to note that we conducted hundreds of such calculations in our work on first-strike stability—armed only with handheld calculators. I believe that our reliance on hand calculations allowed maximum transparency and greatly advanced our understanding of the nuances of first-strike stability. It allowed us to arrive at conclusions that would likely have eluded us had we concentrated on building, running, and

---

[26] See Kent, DeValk, and Warner, 1984.

[27] In the early 1990s, several analysts, among them Dr. Jerry Bracken, extended the methodology to analyze multisided first-strike stability.

debugging a computer model. During our interactions with some analysts using models (like the Arsenal Exchange Model) to assess forces on the basis of stability, we were often shocked by their lack of insight and the rather pedestrian conclusions they put forward. Only much later, after our work was nearly complete, did we duplicate our hand calculations in an electronic spreadsheet to allow us to assess alternative force structures rapidly and repetitively. But it was the process of working through these calculations and results by hand that proved most effective in illuminating the issue of first-strike stability.

CHAPTER THREE

Analysis, Force Planning, and the Paradigm for Modernizing

General Kent returned to Washington from the Air War College in 1957. Other than a one-year stint at Harvard, he never left again. During more than 40 years in and around the Pentagon, he mastered the art of helping high-level decisionmakers understand complex policy problems and evaluate options for addressing them. In this chapter, General Kent offers his views not only on how to structure and conduct analyses but also on how to present one's findings, recruit and train capable analysts, and run an organization whose mission is to develop policy-relevant analyses. He also shares insights about how the services should structure themselves to perform one of their most important roles: developing the operational capabilities combatant commanders need. Although this is a central function of the services under Title 10 of the U.S. Code, parts of the process are surprisingly ad hoc, while others are burdened by layers of bureaucracy. Finally, in this chapter, General Kent revisits the origins of the "strategies-to-tasks" framework, which he devised and which has been used both to identify priority operational needs and to advocate for programs to meet those needs.

On Analysis

Too much has been written by too many on how to do analysis.[1] But too little has actually been accomplished. At the risk of being placed

[1] I thank the editors of *Air and Space Power Journal*, formerly *Air University Review*, for permission to reprint these excerpts from two articles I wrote in 1967 and 1971.

in the first category, I offer some remarks in the hope of enhancing the state of understanding as to how to go about achieving good analysis.

Simply stated, the purpose of an analysis is to provide illumination and visibility—to expose some problem in terms that are as simple as possible. This exposé is used as one of a number of inputs by the decisionmaker. Contrary to popular practice, the primary output of an analysis should not be conclusions and recommendations. Most studies by analysts do have conclusions and recommendations, even though they should not; invariably, whether or not some particular course of action should be followed depends on factors quite beyond those that have been addressed by the analyst. A summary is fine and allowable, but conclusions and recommendations by analysts are, for the most part, neither appropriate nor useful. Drawing conclusions and making recommendations (regarding these types of decisions) are the responsibilities of the decisionmaker and should not be preempted by the analyst.

Under the heading of "summary," one can write quite perceptively, stating that, within the factors we have been able to quantify, if such and such is true, this is the outcome. But, most important, one is not required to go beyond the factors that have been analyzed and make a recommendation that surely is based in part on factors that have not. For example, a particular analysis might demonstrate that a new type of aircraft is far more effective than currently available aircraft in accomplishing certain operational tasks. It does not necessarily follow, and the analyst should not recommend, that the service should procure this new aircraft. That decision must be made on the basis of a wide range of factors, such as overall budget levels, fleet age, and competing operational needs, that lie far beyond what was encompassed by the analysis. Of course, there are the unuseful recommendations. A common one of this type is something like "The subject requires further study." Not only are such statements of little import, but such a conclusion is usually quite obvious without being stated.

Analyses are allegedly undertaken for the purpose of providing illumination. Still, at times, the light has a green tinge, or a deep blue tinge, or a light blue tinge, or a purple tinge. Sometimes the light comes out pure black. Seldom do analysts produce illumination with pure white brilliance. So the decisionmaker becomes wary—as well he or she should—of these biased or shaded illuminations. There must be something wrong when quantification of some particular problem produces radically different results that favor the interests of one service or another. In the blind rush to be effective advocates, analysts enthusiastically engage in practices that border on perjury. The naïve exclaim that the answers appear to have been known ahead of time. The calloused inquire whether there is another way.

There is no easy fix to the problem of parochialism. A common suggestion—in the interest of objective analysis—is to establish joint organizations for analysis or have analyses done by people who are "above service bias." This sounds good, but the theory is better than the practice: Too often it is merely substituting one form of parochialism for another. To be more pointed, the illumination of problems addressed by analyses performed by the services will predictably reflect their own color. The illumination afforded by JCS studies has a way of coming out black because it goes through all the filters. Those by OSD come out purple, which may or may not be a better (or wiser) color than green, deep blue, or light blue. All too often the analyses are conducted in the context of a preconceived position. They become papers for advocacy as distinct from papers for illumination. The quantification is shaped, twisted, and tortured to establish the "validity" of some particular point.

Analyses by OSD and think tanks do not escape this plague for one reason: Their analyses are not usually as subject to critical review by nonbelievers as are analyses from the services. Whatever objectivity is achieved by the services does not necessarily stem from basic purity but rather from fear of rebuttal. One could get a single answer to a particular problem by never having more than one analyst work on the problem. While this would resolve the problem of getting different answers, it would not address the nagging concern about parochialism. Such a measure may clear up the symptom but does not cure the disease.

Aside from bias and preconception, there is another reason analysts get different answers to what seems to be the same problem. Simply put, there is too little discipline in the analysis business. Not all of us handle interactions among forces in the same way. True, different analysts may use the same formula in describing the interaction of a bomb against a target, but once you get much beyond this first, most basic stage, there is little agreement.

Many people who call themselves analysts are really calculators. They spend more time having calculations done on a computer than they spend in analyzing the results. They are expanders rather than distillers. They can be identified easily by the pride that they exude when they present a decisionmaker with a detailed study and announce how many hours it took to generate all this material with a high-speed computer.

The difference between an analyst and a calculator can be grasped by considering the following example: If an analyst is asked what is the effect of reducing, by a factor of two, the CEP of a missile in attacking hard targets, he or she will derive simple statements such as, "If the CEP is halved, it takes only one-fourth as many missiles to have a certain assurance (damage expectancy) of killing a certain number of targets." Further, he or she will add that the ratio of four to one is independent of the hardness of the targets being attacked, the absolute value of the CEP, the assurance desired, and the number of targets.

The calculator, by contrast, will run a number of war games and, if he or she is persistent, may discover that the ratio of missiles required for the lower CEP is about 3.948 for some particular set of circumstances. But rarely will calculations expose universal truths. If at all possible, the analyst should reduce (collapse) the problem to a simple formula or set of formulas with graphs or tables.

Notice that I use the word *analyst* rather than *mathematician*. Granted that, to be an analyst, knowledge of differential calculus is useful, if not essential. But the big task is figuring out how to develop a construct of the problem so that there is something to differentiate in the first place. Mathematicians who can manipulate formulas in a mechanical sense are as easy to come by as the calculators, but analysts are not. As a matter of fact, I have come to the conclusion that the makings of a good analyst are more apt to be found in a lawyer who has a smattering of mathematics experience than in a mathematician who is a calculator rather than a thinker. Since lawyers are generally not particularly well schooled in calculating, they are forced to think and reason, and this is a very good thing.

Analysts should be recruited because they have the talent to dissect problems—to collapse seemingly complicated problems into much simpler terms. They are to be graded on impeccable logic and correct arithmetic. They are to be graded as well on how elegantly and simply they were able to "model" some problem. And I use the verb "model" here not in the conventional sense of setting up a computer program to run calculations but rather in the sense of creating a conceptual depiction of key interactions in a way that permits rigorous quantitative comparisons to be made among different cases without doing violence to the essential aspects of the reality being examined.

One recruits such people from those who have been educated in economics, logic, and mathematics. One looks for people who have exhibited an uncommon ability to think and explain. Position-takers, on the other hand, are graded on how many times their position is accepted by the boss. Position-takers are recruited from people who have good background experience and possess intangibles such as "mature judgment." Of course, the respective talents of these two different groups are not necessarily mutually exclusive. But they are not necessarily coupled either.

The best education for an analyst is in the school of doing. This presupposes that the person involved is alert, curious, and eager to work. Further, he or she should feel somewhat at home with integral and differential calculus. But, given this background, the best way to become an analyst—if there is indeed such a type as distinct from other people—is to work on problems. Guidance and assistance from someone who has been through similar studies are quite helpful. But, ultimately, good studies are produced by hard and earnest work. They are the result of going over and over and over and over the same problem with a view to reducing and collapsing it, on the one hand, and providing illumination and visibility, on the other.

Probably the best procedure for a student who is preparing to embark on a career in analysis is to review carefully the analytical techniques that were used to good effect in analyses already accomplished. With luck, one of these techniques might apply to the problem at hand.

In my view, the courses on analysis now being taught at various schools place far too much emphasis on statistical theory and too little on case histories. The emphasis should be on how to think about problems so as to simplify them. I know of no better way to do this than to review what has been demonstrated in the past. Unfortunately, the textbook I am talking about has yet to be written, but a noble beginning would be for someone to publish a compendium demonstrating the better techniques that have been used to date.

Portraying and presenting one's findings clearly can be as important as generating them. Too many times the results of what was potentially a good analysis go down the drain because of poor presentation. This goes for both oral and written efforts. I have a theory that each listener or reader has a threshold for "naggers." *Naggers* are things that he or she does not understand. When the threshold is exceeded, he or she quits listening or reading. The naggers can come in several forms, all used by presenters at some time or another, for one reason or another. A common practice is to fail to delineate clearly how a particular curve was derived. Now, the ingredients for deriving the curve are almost always contained (submerged) somewhere in the prose—a little clue here and another clue there—and a determined sleuth can finally piece the whole thing together. The trouble is that most readers are not that determined, and they give up. The credibility of a curve will not be established with those who matter unless they can reproduce, at least in concept, the points on the curve. Without establishing credibility, one has little or no chance of making any of the points one may have had in mind. The day has long since passed when one could get away with pronouncements such as "Since the bar for System A is longer than the bar for System B, we should buy System A." The fact that the bar for one system is longer is of little import unless the decisionmaker believes the analysis, and this belief can be established only by the clearest exposition. Sometimes the lack of clear exposition is purposeful, an effort to submerge some awkward or shaky input. To think that such a practice can possibly pay off borders on idiocy.

Packaging is important in many endeavors; the business of analysis is no exception. If the analyst invents new terms, all right; but he or she should announce up front what is being done and then be consistent, not reinvent a new vernacular on each page and chart. There are no problems in this respect that "murder sessions" and good editing will not cure. Some decisionmakers are reluctant to admit that they do not understand some chart, particularly when everyone else in the room has assumed a knowing look. But if the analyst-briefer's charts display strange abbreviations designed primarily to cue the briefer on what to talk about next, the decisionmaker may get tired of reading them, since he or she gets no message. The worst fate of an analyst is not to be contested, but to be ignored.

Other times, a lack of clear exposition in an oral presentation stems simply from a well-known and prosaic disease: The briefer does not know his or her subject. Someone else provided the curves. The briefer thought he or she understood them, and ostensibly did, until someone asked a question that was not in the script. Oral presentations also suffer many times from a plethora of charts and a paucity of message. The best illumination stems from a few charts that are well explained.

What are the fixes for these ills? They can be summed up in one word: discipline. Air Force personnel should apply the same rigid discipline to analysis that they do to flying an airplane. The accident rate for analysis is quite high. However, these accidents are, for the most part, not as dramatic and personal as aircraft accidents, and consequently, there is no concerted campaign to reduce the rate.

If nothing else, poor analyses reflect adversely on our professional image. But how do you apply discipline? You go over and over and over each bit of logic and each calculation. By *you*, I mean you. If it is your study, you should be able to reproduce, when called upon, any number in the study in a reasonable time and without too much fumbling. You only really understand something after you have made the calculations yourself. If the study is so complex that you feel you simply cannot master the calculations, one of two things (or both) is wrong: Either the study is too complex, or you are a poor analyst and should take up another pursuit. A general rule regarding simplicity is that "even generals must be able to understand it." Many of the top people in DoD make it a point to understand important analyses in considerable detail. Rather awkward situations are created when the analyst and intervening echelons do not do likewise in advance.

After all, simplicity, in the interest of illumination, is what we are after. If you are asked to explain something, and in lieu of a direct answer, you start out with "Well, it's rather complicated," you are losing altitude fast. Ambiguous answers to direct questions have the same fleeting value as the air above you and the runway behind you. So the first part of discipline is to keep it simple. The second part of discipline is to explain fully and clearly. In a written text, for each graph or table, one should have a facing page (or pages) with three sections: one that describes the purpose of the graph; one that describes the basis for the computations, including all values for inputs and assumptions; and one that tells the reader what message is to be derived from the graph or table. Now, if you find it trying or difficult to write the third section, you might give serious consideration to omitting the graph in the first place. Exercising this discipline in a written report also helps any oral presentation, particularly if the writer is also the presenter—and he or she should be. At the risk of being repetitious: You learn the details only by getting your hands dirty in the actual derivation of the report. A deep-tanned colonel with a resonant voice and a low golf score is no substitute for a pale-skinned major who has not had much sunshine because he is the one who has been doing the dirty work.

Analyses should be conducted jointly by analysts who are inclined to different positions. The steps are straightforward. First, agree on the relevant measures of merit; second, agree on the factors that affect these measures of merit; third, agree on the form of the equations that describe exactly how each measure of merit is affected by each factor; fourth, agree on the numbers—on what values to assign to the inputs (the factors); and finally, agree on how to present the results.

There should be agreement at least through the third step. This allows the calculations to be made. Agreement may not be reached on the values of all the inputs, but the results for different values can be shown. "If assumption X is used, this is the answer; alternatively, if assumption Y is used, this is the answer." In this way, it is crystal clear why different results are achieved—different inputs were used. At present, all too often, it is not known why different results are attained—

one group used Code 99 and the other 007, and they talked right past each other.

In closing, I would like to go back to the matter of whether or not to include conclusions and recommendations in analyses. Decisionmakers, with good reason, often feel that their responsibilities are being eroded in some fashion or another by the analysts. This concern sometimes takes the form of "These studies will never take the place of military judgment." The rejoinder by the analyst to this charge should be "Sir, my hope is that a decision by you, based on your excellent judgment aided by my elegant analysis, will be better than a decision based on your judgment alone. I can hardly believe the aid afforded by my analysis could be counterproductive." But being confident that an analysis is not counterproductive is oftentimes not that straightforward—particularly if conclusions and recommendations are included. Besides that, the analyst cannot make such a statement in the first place unless he or she has been careful not to preempt the decisionmaker.

As stated earlier, the prime purpose of a defense analysis often has to do with providing illumination on the utility of a particular weapon system or piece of equipment. This illumination provides the basis for the Air Force proposing (or not proposing) that the system should be developed or procured; that is, its utility is such that DoD should (or should not) spend money and resources to acquire it.

Actions to gain resources center on proposals to implement new concepts. To paraphrase Shakespeare, the proposal's the thing wherewith we'll tap the coffers of the king. The central question is whether or not a proposed concept should be implemented. Analysis, hopefully, provides added insight into this all-important question.

Running Air Force Studies and Analysis

My General Approach and the "Learn to Think" Mandate
In early 1968, I held the position of Development Planning in Air Force Systems Command (AFSC) at Andrews AFB. One day, I received a call from the office of the Air Force Chief of Staff, Gen John McConnell. The chief wanted to see me at two o'clock that afternoon. I

felt that I should inform my boss, Gen James Ferguson, head of AFSC, of this development, so I asked the caller, "What is the subject?" The lieutenant colonel on the other end said, "I will check." He called right back and related, "The subject is the chief wants to see you in his office, alone, at 2."

"I will be there," I replied.

I entered the chief's office. "Sit down. I want to talk to you, boy." (General McConnell called me "boy" from the very start, even though I was a two-star general.)

"I need someone on my staff reporting directly to me who thinks hard and independently and will tell me what I should hear and not what he thinks I want to hear. I have been watching you, and I think I can count on you. You will assume the position of head of AFSA next Monday."

Wow. What an opportunity. I had the ear of the chief and was head of an organization to help me think hard and independently. For me, this would be a dream job. I had spent my career attempting to understand and communicate more effectively about matters relating to the defense of the United States, and now I was being given an entire organization to enlist in this effort.

This episode had a strong influence on how I ran AFSA. General McConnell wanted insight; he could not have cared less about models. If a computer model were a means to gain insight, so be it. But analysis would not be done for its own sake; the end product was to provide insight. The general really did not care about analysis per se. But he did like what analysis could do for him.

A short time after I took over at AFSA, I received a call from Lt Gen Robert Dixon. He was then serving on the Air Staff as Deputy Chief of Staff for Personnel. I knew General Dixon since we had both been colonels earlier in the office of the Deputy Chief of Staff for Plans and Programs, both working for Maj Gen Glen Martin, the Director of Plans. General Dixon told me that, within reason, I could have any officer I wanted. "Go out and recruit the best," he said. I took his statement to heart and it served me well. Allowing me to recruit good people soon made AFSA an elite organization.

After I arrived at AFSA, one of my first moves was to reduce the emphasis on computer models. At the same time, I directed that ponderous studies that would not yield insights for a year or two be either reoriented or cancelled. I also installed a process of intense and critical review of all papers before they were presented to the outside world. These directives did not sit well with some of the analysts. My deputy told me that morale was sinking and that there was a large measure of discontent.

Accordingly, I called a meeting of the entire organization (maybe 40 people) and informed all assembled that I was not about to change my ways. "If any of you feel you cannot (or do not desire to) work for me, all right. Let it be known and you will be transferred without prejudice." A little more than one third submitted a "no." Fortunately, none of the officers I wanted to keep were on the list. I had lucked out. I now had all those empty slots for which I could recruit good people. A guiding principle for four years was to recruit good people, set them to work on problems that were likely to come before the chief, and then stand back.

I took recruiting very seriously. For example, each year I would go to the Air Force Institute of Technology at Wright Field and interview their top four graduates in that year. From these interviews, I would select one or two—usually lieutenants. So I was getting the cream of the crop of young officers.

One of my approaches to providing critical review was to hold roundtable discussions, or murder sessions, on important and complex problems. There was no rank at the discussions. Lieutenants and captains had as much right to the floor as the majors and colonels. If someone said something that was incorrect, that statement was to be challenged immediately by someone. From time to time, I even engaged in entrapment of a sort. I would read a statement that contained a statement of doubtful validity. If no one corrected me, I would remind them that we were gathered here to learn to think about this problem. "Captain so-and-so," I would say, "were you asleep, or were you just being overly polite? I hope you know better."

In later years, many officers who went on to higher positions told me that the one thing they learned at AFSA was how to think. They continued the idea of murder sessions, in which you are obliged to think more sharply and insightfully about a problem to hold your own in heated discussions, first in house and later in the outside world. It was not entirely a democracy. At the end of a session, I would state the construct of the problem at hand; unless someone, within a week, presented a compelling argument for a better construct, then that way of thinking would stand.

Unfortunately, such a practice of rousing murder sessions is not alive and well today. The person in charge has to be able to invest considerable time and effort, and many leaders are simply spread too thin to engage in such practices. Others are reluctant to do so, lest their lack of insight become apparent to the staff.

The success of AFSA was due primarily to the talents of its people. My contribution, whatever it was, centered on the discipline of thinking and in discourse and, at times, motivation: I was able to put the right people to work on the right problems, and at the right time.

The individual pieces of work that made a difference in the business of the Air Force are chronicled in other chapters of this book.

On Recruiting

One of my greatest and most fortunate successes in recruiting was to have Col Jasper Welch in my organization. He had done brilliant work on the damage-limiting study in 1964 (see "Limiting Damage to the United States," pp. 43–50) as a major, and now he was a colonel. I hoped to have him serve as AFSA's "chief military analyst." His skills knew no bounds. He could program a computer to do his bidding better than the full-time programmers could. This was no mean feat in the 1960s. Computers were not user-friendly. He also knew as much about scientific matters as the best of scientists.

I worked hard to recruit him. He was about to leave the Industrial College of the Armed Forces at Ft. McNair. He was scheduled, first, to work for Lt Gen Otto Glasser, Deputy Chief of Staff for Research and Development. That was changed to an assignment to work for me—no mean feat in itself—only to have the Secretary of the Air Force, Dr. John McLucas, make a request for Colonel Welch to work for him. After four letters to the secretary in rebuttal, he graciously gave up. My

argument was simple: If Colonel Welch were in AFSA, he would educate others; we could leverage his talents. And that is how it worked,
much to the benefit of the Air Force.

The Paradigm for Promoting Innovation and Modernizing the Operational Capabilities of the Force

The armed forces of the United States have, over the years, brought forth
some remarkable innovations: radars that can detect targets at night
and through clouds, low-observable (stealth) aircraft, and precision-
guided weapons. These new technologies have transformed military
operations, allowing air forces to achieve objectives far more rapidly,
with less cost and risk, and with less collateral damage than was the
case just a generation ago. So DoD (and the Air Force in particular)
is indeed capable of innovation. But, too often, innovations occur in
spite of, rather than because of, the defense acquisition system. As that
system has become more elaborate, the time it takes to bring new platforms to the field has grown, the costs of systems have increased, and
the sense has become pervasive that we are delaying (or missing) opportunities to provide operators with better capabilities.

Chapters Four and Five of this volume provide examples of the
ways in which an obsession with square-filling and "requirements" can
become a roadblock to efforts aimed at modernizing capabilities. But
the tyranny of the requirements process is only one manifestation of
the broader problem of promoting innovation within DoD. True innovation involves discovery—the discovery of new ways to accomplish
important operational tasks. By its very nature, this process is more art
than science. In many instances, the solution to an important operational problem is not so much a matter of inventing new technology
as one of creatively employing or adapting technologies that, in one
form or another, already exist. I call this creative process of marry-
ing operational need to technological opportunity *concept development.*
Concept development lies at the heart of modernization. And modernization lies at the heart of a military service's responsibilities. Yet the
services, by and large, have been content to leave concept development

to chance. Most of the time, it has been an ad hoc process left to industry or simply to serendipity.

Another barrier to innovation is the "defense acquisition system."[2] This term has been applied to the organizations and procedures that are supposed to govern the development, testing, procurement, and management of new military systems. The regulations governing the defense acquisition system have been revised every few years, generally under the 5000-series of DoD directives.[3] But the general characteristics of the system have remained more or less constant:

- a set of "milestones" that must be accomplished and certified before services are permitted to proceed with the development of new concepts
- a tendency to conflate statements of operational need with potential hardware-oriented solutions
- a stultifying inclination to impose centralized control over efforts to explore new concepts for accomplishing operational tasks.

Throughout my career, I sought to overcome both the laissez-faire attitude that the Air Force (and its sister services) have toward concept development and the pernicious tendency of the requirements and acquisition bureaucracies to impede innovation. In 2003, David Ochmanek and I published a short report that sought to help people interested in modernization to carry on these fights.[4] The remainder of this section is a summary of the main points of that report.

A Framework for Modernizing

When airmen speak about the conduct of military operations, they emphasize the importance of "centralized planning and decentralized

---

[2]   Even the name is wrong. Taken literally, *acquisition* should apply only to buying things. The processes of defining and developing those things and, more importantly, determining the operational concepts that will govern their employment, should be referred to as *modernization*.

[3]   See, for example, Department of Defense Directive 5000.1, *The Defense Acquisition System*, May 12, 2003.

[4]   Glenn A. Kent and David A. Ochmanek, *A Framework for Modernization Within the United States Air Force*, Santa Monica, Calif.: RAND Corporation, MR-1706-AF, 2003.

execution." By *centralized planning* they mean that, in any major operation, the commander must formulate a plan that makes best use of the assets he or she has available and that governs the employment of those assets. The *air tasking order* is typically the vehicle for operationalizing the commander's current plan and communicating it to the units that will execute it. By *decentralized execution*, they mean that the commander's responsibility does not extend to planning and flying each sortie. Operational units do that. Our framework for modernizing shares this basic philosophy. We believe that our development efforts suffer because they lack clear, consistent guidance on what sorts of operational capabilities should be given highest priority. It is a responsibility of each service headquarters to provide this guidance. But headquarters should not aspire to micromanage the creative process of concept development. This is best left to multidisciplinary teams devoted to solving the sorts of operational problems that the leadership has identified as being most urgent.

Functionally speaking, seven principal "actors" are involved in the modernization process within a service:

- *The definer's* chief role is to frame a finite set of high-priority operational challenges (or operational requirements) that the Air Force will strive to meet.
- *Proponents* define new concepts of employment (CONEMPs). A CONEMP is a concept for achieving a particular operational objective. Each proponent is responsible for monitoring and assessing the Air Force's capabilities to achieve a related set of operational objectives. The proponents also seek to ensure that adequate resources are allocated within the Air Force to sustain and advance "their" set of operational capabilities.
- *Conceivers* formulate; define; and, when appropriate, demonstrate new concepts of execution (CONEXes). A CONEX is an end-to-end concept for accomplishing a particular operational task.
- *Independent evaluators* advise the Secretary of the Air Force and the Chief of Staff on the merit of proposed new concepts.

- *Programmers* estimate the cost of proposed concepts and suggest ways for balancing resources across all the activities that the Air Force carries out.
- *Providers* supply *capabilities* (not "forces") to combatant commanders by organizing, equipping, and training units to accomplish CONEXes and achieve CONEMPs; the acquisition of new platforms, weapons, and support systems falls under this rubric.
- *The Secretary of the Air Force* and *the Chief of Staff* preside over the entire process and render decisions at key points. These decision points include (1) the issuance of an approved list of operational challenges, (2) the choice of whether to pursue a concept proposed by the proponents, and (3) advocating that concept to OSD to gain the resources needed to implement the concept.

The system for spurring and managing innovation within the Air Force is established by defining the responsibilities of each of the aforementioned actors and the relationships among them. Figure 3.1 illustrates the primary functions of each of these seven actors and the interactions among them. In the figure, darker ellipses indicate entities within the Air Force; lighter ones designate entities outside of the Air Force. I followed this concept or model for many years, to good effect, in governing the efforts of the Air Force to modernize the operational capabilities it provides to combatant commanders.

Strategic Planners and Definers. The process starts in the upper left corner of Figure 3.1 with the strategic planners. These planners reside in OSD and in the Joint Staff. They reside as well among the National Security Council staff and in various think tanks. The output of these planners is a series of statements regarding the future operating environment, the possible missions of the U.S. armed forces, and the types of capabilities that the planners believe will be most relevant to future military operations.

The judgments of these planners, being framed at a high level of generality, are not always directly useful in defining the types of operational capabilities the Air Force intends to provide. For that reason, our model also features a definer within the Air Force, who determines

Figure 3.1
The Framework for Modernizing: A Service Perspective

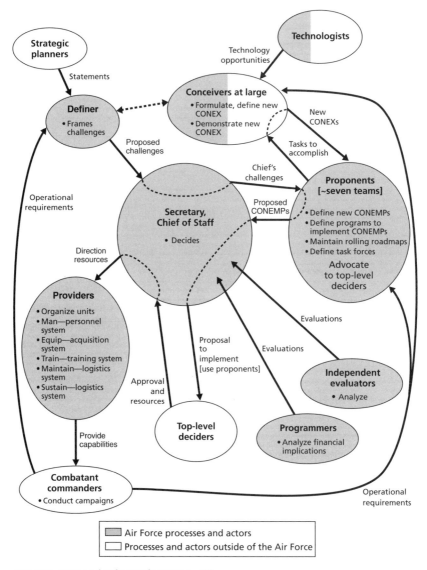

SOURCE: Kent and Ochmanek, 2003, p. xiii.
RAND OP223-3.1

what these statements mean in terms of the types of operational capabilities to be provided by the Air Force.

For example, translating such terms as *rapid decisive operations*, *network-centric warfare*, and *precision strike* into statements of challenges on which the Air Force will focus is not a straightforward proposition. That is why the definer is shown as a principal actor. Surveying the changing security environment, participating in the development of defense strategy, and helping to identify looming shortfalls in the capabilities of joint forces is a full-time job for a creative major general.

The output of the service's definer, then, is a set of "challenges." These challenges are statements that, when approved by the secretary and the Chief of Staff of the service, task the corporate Air Force to focus on developing operational capabilities appropriate to meeting each challenge. As such, these challenges can be considered *operational requirements*. This being the case, it is important that the challenges be derived from an examination of joint campaigns and that they be phrased in terms of operational objectives to be achieved by joint forces. The definer, then, should identify a set of challenges that joint forces face, which the capabilities provided by Air Force forces can help to address. For this reason, Figure 3.1 shows the combatant commanders communicating their various operational requirements to the definers, the proponents, and the conceivers.

The set of challenges developed by the definer is presented to the secretary and the Chief of Staff for their review; adjustment; and, finally, approval.

Proponents. The next principal actors in this process are the proponents. Their principal function is to develop, in response to the operational challenges, new CONEMPs—new concepts that will enable Air Force forces, operating in conjunction with the forces of other services, to achieve important operational objectives even in the face of enemy resistance. Because modernization is such an important function of the service, several senior officers should be vested with the responsibility for ensuring that the service is pushing the state of the art in developing important operational capabilities.

The principal input to each team of proponents is one (or more) of the chief's challenges. We proposed in the report that an Air Force proponent be placed in charge of seven teams. Each team would be challenged to define new CONEMPs,[5] so that the Air Force can provide improved capabilities to fulfill this operational requirement—now and especially in the future. The seven teams would be constituted as follows:

- *Team 1: Gain freedom to operate.* This encompasses efforts to establish access in theaters of operations, to gain air superiority and space superiority over the enemy, and to sustain a high tempo of operations at bases in the theater despite countervailing actions by the enemy.
- *Team 2: Provide control of the operation of forces.*
- *Team 3: Provide strategic airlift.*
- *Team 4: Fight and gain the effects desired in conflicts.*
- *Team 5: Protect the homeland.*
- *Team 6: Conduct global strikes.*
- *Team 7: Conduct other operations.* This includes maintaining global awareness, providing a stabilizing presence in key regions, and providing humanitarian relief.[6]

Conceivers. There is another input to the proponents. This input, generally in the form of new CONEXes to accomplish military tasks, stems from the conceivers. The conceivers, in turn, have two inputs: (1) a knowledge of existing and emerging technologies and (2) statements from the proponents as to the military tasks that merit the greatest emphasis. In the presence of these two inputs, the conceivers "connect the dots." They define concepts (the CONEXes) for accom-

---

[5]   *CONEMP*, as defined here, refers to the means for achieving a stated operational objective or for conducting a major operation within a campaign. For example, a CONEMP would be defined for how forces should defeat the enemy's air defense and so gain freedom of action over that country. *CONEXes* are the means for accomplishing stated military tasks. In this example, tasks associated with the objective of defeating the enemy's air defenses might include destroying air-defense control centers, destroying or suppressing SAM tracking and guidance radars, and shooting down enemy fighter aircraft in flight.

[6]   The Air Staff instituted groups similar to these in 2003. They designated the head of each one a "concept of operations champion."

plishing stated military tasks and, when appropriate, conduct field tests to demonstrate promising new concepts.

The output of the conceivers is an evolving portfolio of CONEXes—a portfolio from which each of the proponents can pick and choose.[7]

Back to the Proponents. In the presence of new CONEXes generated by the conceivers, the proponents formulate, define, and demonstrate new concepts (the CONEMPs) for achieving stated operational objectives. Episodically, the proponents communicate to the secretary and the chief that they approve the proposed concept and that they seek authority and resources from OSD to implement the concept. There is a role for people in the "joint world" to participate in the development of new CONEMPs. Personnel at Joint Forces Command, on the Joint Staff, at the headquarters of the combatant commands, and elsewhere may devise creative ways to put together the "piece parts" (CONEXes and systems) developed by the services into CONEMPs (at the operational level) that meet the needs of combatant commanders. But generally, operators and technologists in the services will be best positioned to formulate new CONEXes (at the tactical, or task, level).

The Secretary and the Chief of Staff. At this point, the leaders of the Air Force decide whether or not to approve the concept as proposed (or amended). If they elect to proceed, they propose to the top-level deciders (the Secretary of Defense and his principal advisors) that the concept be implemented and that resources be allocated to develop the requisite hardware, facilities, and other assets required to make the concept a reality and, when appropriate, to organize, equip, and train new units.

The Independent Evaluators. The secretary and the Chief of Staff, in their deliberations regarding what to propose to the top-level deciders, rely heavily on the independent evaluators. The proponents, of course, conduct their own evaluations. Their evaluations focus mainly

---

[7] For more on the role and process of concept development, including a description of the work of an ad hoc concept development group, see John Birkler, C. Richard Neu, and Glenn A. Kent, *Gaining New Military Capability: An Experiment in Concept Development*, Santa Monica, Calif.: RAND Corporation, MR-912-OSD, 1998.

on a stated operational objective. Specifically, each proponent seeks to determine which among several candidate CONEMPs will best achieve the stated operational capability for which he or she is responsible.

The independent evaluators also address the issue at this level. However, they have a higher-level focus as well. More specifically, they address the question of how resources should be allocated among the full range of CONEMPs being considered and proposed by the proponents. The independent evaluators, then, strive to shed light on the question of whether or not to implement the proposed CONEMP in the presence of competing demands for resources.

Financial Managers and Programmers. Over time, the proponents will be advocating the implementation of a number of CONEMPs. Before deciding yea or nay, the secretary and the Chief of Staff must hear from the independent evaluators regarding the value of each concept in terms of operational capabilities. They must also hear from the financial manager and the programmer as to financial implications. The financial manager and programmer must first estimate the full costs of adapting a new CONEMP and determine the resources that would be required on a year-by-year basis. They determine whether the concept would fit (or not fit) into the overall Air Force budget, again year by year. In this connection, the programmer is obliged to inform the leadership if the proponent has significantly underestimated the cost to implement the concept.

The Top-Level Deciders. Once a proposed CONEMP has made its way through each of these actors, the Air Force is ready to approach the top-level deciders. Here, the top-level deciders comprise the Secretary of Defense and his principal advisors, chiefly the Under Secretary for Acquisition, Technology, and Logistics. The input to the top-level deciders is "proposals"; their output, hopefully, is "approval" and "allocations." The top-level deciders base their decision on the same criteria as the Air Force's Secretary and the Chief of Staff:

- Is the CONEMP technically feasible?
- Is the CONEMP operationally viable?
- Is the capability provided by the CONEMP relevant in conducting important military operations?

- Is the CONEMP consistent with policy and other political constraints?
- How well does this CONEMP measure up against competing CONEMPs on the basis of marginal return?
- Is the acquisition program to acquire the systems associated with the CONEMP executable?

Based on the answers to these criteria, the top-level deciders make a decision to implement (or not implement) the CONEMP.

A service can probably be counted on to propose the best CONEMP within the realm of its medium of operations (air, land, sea, space) to achieve a stated type of operational capability. The expertise is in the service, and there is every incentive to propose the best CONEMP. On the other hand, a particular service can hardly be an unbiased witness to the issue of whether resources should be allocated to implement the particular CONEMP it has proposed, as opposed to CONEMPs proposed by other services. Accordingly, top-level deciders will rely on their own independent evaluators to assess the value of proposed CONEMPs and their likely cost.

Once the top-level deciders have rendered their decision to implement a concept and inform the secretary and the Chief of Staff of this decision, the secretary and the Chief of Staff issue direction to the providers and allocate resources accordingly.

The Providers. The input to the providers is direction to proceed. That is, they are directed to provide the stated operational capability by implementing the proposed CONEMP within a given level of resources and according to a certain schedule. Implementing a new concept involves (or could involve) several functions:

- organizing units
- manning these units with trained personnel
- equipping these units with systems, weapons, and other hardware
- training the operators in the units
- maintaining these units in peacetime
- sustaining these units in combat operations.

As stated earlier, the acquisition system equips units with systems. Because Title 10 prohibits the secretary of the service from delegating the oversight of system acquisition to the service chief, the secretary and his staff play the leading role in this important function.

In closing, the output of the provider is operational capabilities that are available to the combatant commands. In this way, the Air Force carries out its responsibility, so succinctly stated in Title 10 of the U.S. Code, "to fulfill the current and future operational requirements of the unified and specified combatant commands."[8]

The process described here is, admittedly, a stylized picture of how the modernization process should unfold. But it is useful both as a means of orienting oneself in the chaotic, bureaucratized system that actually exists, and as a model toward which the leadership of the service and DoD as a whole could aspire.

Strategies to Tasks: A Construct for Advocating New Concepts

Around 1970, the advocacy of Air Force programs to acquire new systems became the center of attention for Gen John D. Ryan, then Chief of Staff of the Air Force. The Air Force had submitted to the Senate Armed Services Committee less-than-sterling papers explaining the rationale for three of its key modernization programs:

- the Airborne Warning and Control System (AWACS)
- the F-X (which would become the F-15)
- a lot buy of the C-5A.

The chairman of the committee deleted all three of these programs from the defense authorization bill—all in one day.

Needless to say, this action got the full attention of the Air Force Chief of Staff and, by extension, the rest of the Air Staff. General

---

[8] U.S. Code, Title 10, Armed Forces, Chapter 803, Department of the Air Force, January 19, 2004.

Ryan's reaction to this turn of events and the approach we might take toward preventing a recurrence are instructive for those involved in force modernization.

Strategies to Tasks Debuts

The origin of the crisis was an effort by Sen. William Proxmire (D-WI) to eliminate funding for the development of the E-3 AWACS. In response, Sen. Barry Goldwater (R-AZ) asked the Air Staff for a new paper about the AWACS and its capabilities, a paper that he could use on the floor of the Senate to rebut Senator Proxmire's arguments. The paper, which was prepared by an action officer on the Air Staff, arrived quite late. In fact, it was delivered to the senator by a staff member on the Senate floor just prior to the start of proceedings. Moreover, Senator Goldwater found the content of the paper to be quite uninspiring: It contained mostly technical and programmatic data about the performance specifications of the system and projected development milestones, with very little information about the operational utility of the system. Goldwater's judgment was that, armed with what the Air Force had provided him, he would be unable to make a persuasive case for the program.

When informed of this, Sen. John Stennis (D-MS), chairman of the Senate Armed Services Committee, employed some deft parliamentary maneuvering and postponed deliberations about the AWACS program until another day. Needless to say, both Senators Stennis and Goldwater were quite unhappy about the poor support they had received from the Air Staff. Senator Stennis had no intention of giving his rival, Senator Proxmire, a chance for an easy win on a major defense program. To deny Senator Proxmire such an opportunity, Senator Stennis deleted the AWACS from the bill. For good measure, he also deleted the F-X and a lot buy of C-5A, as Senator Proxmire had amendments pending to delete both of these programs.

Senator Goldwater then penned an angry note to General Ryan. The note, which was handwritten on a five-by-seven–inch piece of paper, arrived around midday on a Friday. It stated, "L and L [Legislative Liaison] means 'Late and Lousy.' How can I support your programs with papers like this? Get your . . . act together."

In response to Goldwater's note, General Ryan held an impromptu meeting of the key officers on the Air Staff. The outcome of that meeting, which featured considerable and graphic language by the chief, was that certain generals were to return to his office the next morning prepared to define an approach to fix the problem of incompetent advocacy of our key programs.

The Saturday meeting opened promptly at 0800. All the "rank" in the Air Staff was there. In an effort to avoid fire, I took a seat toward the far end of the table. Three different lieutenant generals offered their ideas. General Ryan responded to their statements with comments such as "That misses the mark," and "That is worse than the last thing you just said."

Suddenly General Ryan pointed directly at me. "I did not invite you here to sit this out. You usually talk too much. Speak up! Say something!" It was clear at that point that Plan A—lying low to avoid fire—was not working. On to Plan B. "Sir," I said, "I have withheld comment on the previous proposals since they are centered on putting 'Band-Aids' on a basically flawed system.[9] Advocacy must be operational and by the operators, not by the technical people from the SPO [system program office]. We must also prepare our advocacy for programs well in advance of the 'crunch point' in Congress. A group of our best colonels should be established to write these papers. The papers should focus on operational matters and describe the ways in which the operational capabilities provided by the system fit into the big picture. The paper should be 90-percent done before the crisis appears."

At this, General Ryan gave a brief nod and departed the meeting. His plane was waiting at Andrews to take him to Vietnam. After the regular morning staff meeting the following Monday, Gen J. C. Myer, the Vice Chief of Staff, asked me to stay behind. "I received a call from General Ryan on Sunday from Hawaii," he said. "He was impressed by what you said. In fact, he was so impressed that you have the job. According to your own deathless words, you are to form a group of elite colonels. Your first job is to get the three 'lost' programs back into the bill."

---

[9]   Band-Aid® is a registered trademark of Johnson and Johnson Consumer Companies.

I had expected this outcome and went to work right away. I asked Col Jasper Welch, who worked for me, to head the group. The group included Lt Col Larry Welch (later the Air Force Chief of Staff). All the others who were in the group (five or six total) later became generals.

The approach we took to advocacy eventually came to be known as *strategies to tasks*. Central to this construct is to link the "system" at hand to the larger picture—the military strategy that U.S. forces have been called upon to execute. The conceptual hierarchy is as follows:

- *national security objectives* to attain to secure U.S. interests
- *strategies* to carry out to attain the stated national security objectives
- *operational objectives* to achieve that implement the strategies
- *military tasks* to accomplish to achieve the operational objectives
- *systems* to execute the "operational concept" to accomplish the various military tasks.

For example,
- We are engaged in a Cold War with the Soviets. One *national security objective* of that war is to deter and defeat Soviet aggression on the "Central Front" in Europe.
- Our *strategy* with regard to that national security objective is a forward defense. We, the allies, will undertake to halt any Soviet invasion as far forward as possible—hopefully at (or near) the intra-German border.
- One *operational objective* to achieve, attendant to carrying out that strategy, is to delay, damage, or destroy Soviet follow-on forces as they make their way across western Poland and East Germany.
- One *military task* to accomplish to achieve this stated operational objective is to drop bridges.
- Our *operational concept* to drop bridges is to equip our attack aircraft with laser-guided bombs.

Note that the construct is quite operational until we arrive at the level of tasks to accomplish. Systems are not named until this level. Even at this level, the focus is on the operational concept for finding, identifying, engaging, and destroying a class of targets, rather than on the systems per se. The overall construct would have to have coherence from the top down and from the bottom up. Our arguments would go like this: We need this system to execute this operational concept, to accomplish this task, to achieve this operational objective, to carry out this strategy, to attain this national security objective, to prevail in the Cold War. And so was born the concept of *strategies to tasks*, a construct that was enthusiastically endorsed by the leaders of the Air Force—then and later. But we should return to the group.

We wrote several papers according to this construct. Within three weeks, as I recall, we had succeeded. All three programs were back in the bill. The construct of linking each system to a larger and larger picture was compelling. For example, the paper on the AWACS was not titled "AWACS." Rather, it bore the title "Supremacy in the Air on the Central Front." The acronym "AWACS" as such did not appear until the second or third paragraph, and then under the rubric "The Eyes of the Commander and the Controllers." Nowadays, there is a focus on "capabilities-based planning," as though this concept were a recent invention. The construct of strategies to tasks is capabilities-based planning at its best. Strategies to tasks is, of course, about capabilities and more. It defines how specific capabilities fit into a larger overall picture.

The Advocacy of Systems

Senator Proxmire, along with three other senators, had put together a group of analysts capable of quite sophisticated work. One of their analyses concerned a lot buy of the C-5A, then being considered by the Congress. They argued that, rather than acquiring more C-5s, the United States should plan to contract for commercial aircraft in time of emergency to supplement the military airlift fleet. This concept, called the civil reserve air fleet (CRAF), had been in place for some time, and Senator Proxmire and his colleagues sought to expand that fleet on the basis that it would be more "cost-effective" than buying more C-5s.

Our response was not to argue that CRAF was a bad idea but, rather, that CRAF and military airlift aircraft complemented each other and that the key investment decision should focus on determining the proper mix of the two. To demonstrate this, we assumed that the United States contracted for the level of CRAF support called for by the anti–C-5 senators. If the nation did that, how many C-5s would be appropriate in the mixed fleet? We showed that, if the task was to airlift an Army mechanized division from the United States to Europe, the optimum mix actually called for more C-5s than the Air Force was asking for.

The analysis was as follows: Only the C-5 could carry so-called outsized equipment—tanks and so forth. In this construct, the more CRAF aircraft you have, the more C-5s you need for a "balanced deployment." Otherwise, the outsized equipment constitutes the "long pole" in the tent. Their own proposition became a compelling argument for buying more C-5s.

In short, we trumped their analysis. They had not done all the homework they should have. It took two officers almost a month of dedicated effort to get a handle on the number of C-5s required to load the outsized equipment of a mechanized division. But it paid off. The argument was no longer a matter of cost-effectiveness but rather one of an integrated deployment. Even if commercial aircraft were cost-free, we would still need the additional C-5s for a balanced and integral deployment.

Senator Proxmire and company withdrew their amendment, and the Air Force completed its planned procurement of the C-5. General Ryan was quite impressed.

Strategies to Tasks Employed to This Day

The strategies-to-tasks framework has proven to be a flexible tool with a number of applications. Because, at its heart, it embodies a disciplined disaggregation of strategy, it can be used to help analysts and decisionmakers grapple in a systematic way with the demands of particular strategies. After joining RAND in 1983, I published a RAND research note, "Concepts of Operations: A More Coherent Framework for Defense Planning," which detailed the construct of strategies to tasks.[10] This spawned numerous efforts at RAND to apply the framework, and it eventually spread to planning efforts within DoD. During the late 1980s, RAND applied the framework to assessments of U.S. capabilities for defending against Soviet aggression on the Central Front and for deterring nuclear attack on the United States. In March 1993, I presented the framework to the Air Force's Modernization Planning Conference, and the Air Force adopted its own versions of strategies to tasks for its subsequent planning activities. It has also been used to show the contribution of space-based capabilities to terrestrial operations; to support planning and programming in the Special Operations Command; and most recently, to help the Air Force develop plans for the future Iraqi air force.

---

[10] See Glenn A. Kent, "Concepts of Operations: A More Coherent Framework for Defense Planning," Santa Monica, Calif.: RAND Corporation, N-2026-AF, 1983. See also Edward L. Warner III and Glenn A. Kent, "A Framework for Planning the Employment of Air Power in Theater War," Santa Monica, Calif.: RAND Corporation, N-2038, 1984; David E. Thaler, *Strategies to Tasks: A Framework for Linking Means and Ends*, Santa Monica, Calif.: RAND Corporation, MR-300-AF, 1993; and Kent and Ochmanek, 2003.

CHAPTER FOUR

Modernizing Nuclear Forces

As previous chapters show, for much of his career, General Kent was involved in efforts to develop and shape U.S. nuclear forces and the strategies for their deployment and use. In this chapter, he relates his roles in eight specific systems or issues involving these systems. His roles ranged from being a critic of the viability of an area-denial weapon, to being a skeptic about the invulnerability of submarines, to being a proponent of keeping bombers in the triad, to being a developer of the MB-1 rocket. In every instance, General Kent used analysis to shed light on the problem at hand.

Killing the Concept for an Area-Denial Weapon

Even in the early years, I realized that any analysis that addressed an important and controversial issue would surely be subject, sooner or later, to very critical review. So it is advisable to make every effort to ensure that you are right before you expose your "findings" to the cruel outside world. One episode from my career makes this point nicely.

I graduated from the Naval Postgraduate School in 1948 and from the Master's program in Radiological Engineering at the University of California, Berkeley, in 1950. I was then assigned to the Armament Division at the Pentagon. This division was under a two-star general, Maj Gen Don Yates, Director of Research and Development on the Air Force Staff; he was the most demanding and dominating general I ever worked under.

I was promptly granted the proper clearances and was informed about "Project X"—a project that was "close hold." The concept was to take the waste from nuclear reactors, fashion pellets from this waste, put these pellets in fluted spheres (balls) about the size of a softball, put the balls into a dispenser, and dispense these balls over an area.

The idea was to create a barrier. If someone tried to cross the barrier, even in a tank at high speed, the crew would absorb so much radiation as to be incapacitated upon reaching the other side (or soon thereafter). At a more strategic level, creating a barrier on the Central Front would halt tanks in any Soviet invasion. We would create the barrier as soon as possible after the invasion started. We were to dedicate a certain number of B-47s to this purpose and put them on alert.

I was to be the Air Force's action officer on this project—heady stuff for a newly arrived major.

Almost from the start, I had reservations about this concept. Would we be granted permission to use this weapon? The Germans would be concerned about contaminating a large area of their territory. Also, it was a nuclear weapon. What about escalation?

My immediate boss, a colonel who headed the Armament Division, let it be known that these matters were well understood and in good hands. They were above my pay grade. My job was to maintain oversight with regard to technical or engineering matters. He told me to keep out of the other issues.

Still, other concerns about the whole concept began to surface. One day I thought about another problem, perhaps not a technical or engineering problem, but at least a problem at the operational level. The material in these weapons (the radioactive waste) decays. Some time in the future, as much radioactive waste would decay in one day as we would produce on that day, depending on the rate of waste production, which depended on how many reactors would be in operation. Once we determined how much nuclear waste would be available at this equilibrium point, we could calculate the size of the area we could contaminate to prevent passage.

Conducting this analysis was truly beyond my abilities. We needed to know how to calculate the potency of the pellets (a function of how long the pellet had been in the inventory), how many roentgens (a measure of radioactivity) it would take to incapacitate a person, and how many pellets per square mile it would take to create an effective barrier—(a function of how much shielding enemy troops would receive from the structure of the tank and from rough terrain).

Dr. Al Latter of the physics department at RAND had also received the proper clearances to work on the project. I told him of this construct for evaluating the operational effectiveness of this overall concept. He was very interested. He too had misgivings. Accordingly, he organized a considerable effort in the physics department to examine the problem. The results were quite disquieting for advocates of the program: Taking all things into consideration, with the amount of radioactive material we could generate, we could create a barrier probably less than 100 miles in length along the Central Front—only a small segment of that front. In all likelihood, the Soviets would quickly determine the extent of the barrier and go around it. Thus, creating the barrier would do little to halt the overall invasion. We recalled that, in World War II, the Germans went around the Maginot Line, not through it.

I went to Santa Monica and spent two days going over the calculations and assumptions embedded in this analysis. When I returned, I briefed my immediate boss. He was not impressed. He was inclined to dismiss the analysis and keep it under cover, at least until the analysis had been reviewed by others. But General Yates somehow got wind of the analysis and called me to his office. I went over the calculations and dropped the bottom line: We could produce a barrier of less than 100 miles in length. General Yates let this be known to the other generals who maintained oversight of the project.

Unbeknownst to me, one of these other generals immediately chartered a team to critique the analysis. About two weeks later, they briefed the oversight group to the effect that "Major Kent's study" was seriously flawed. I was not invited to this briefing. I had no opportunity to announce that this was a RAND study, not my own.

Later on, General Yates called me to his office. "Let's review the bidding," he said. "You have been assigned here about three months. You made an analysis; I was foolish enough to believe you had a point;

I made it known to the group; they had an expert review it; you are wrong; and I am left hanging out to dry. What do you say to that?"

Well, it seemed my career under General Yates was at an end. "Sir," I said, "the analysis was done by RAND; I will have the people at RAND critique the critique; I am confident we are correct."

General Yates snorted at my statement. "I have no choice but to go to the group and admit that your analysis is fatally flawed," he said. I protested strongly. The general finally relented and gave RAND and me one week in which to critique the critique.

It was a slam dunk. Among other things, their critique contained several careless and serious mistakes—some in simple arithmetic. I let this be known to the general. Now General Yates took another tack, telling me, "Your job is to make your analysis 'stick,' especially with the other generals."

I was not entirely successful in this endeavor, but the pot had been stirred. In due time, the Atomic Energy Division of the Research and Development Board canceled the project. That was the beginning of a fruitful, though often tempestuous, relationship with General Yates from which I profited greatly. So my first venture into analysis on an important issue was a success—mainly because I had gone to great lengths to be sure it was right before venturing forth. The analysis was a success, even though (or in this case because) the patient died.

The B-36 Delivering Megaton Bombs

In the early 1950s, while I was still working for General Yates, I became involved in the effort to equip the new B-36 bomber with hydrogen bombs. The Air Force had announced a "requirement" for the B-36 to be able to deliver a bomb with a yield of 10 megatons—a weapon with a yield more than 800 times that of the weapon that had destroyed Hiroshima. At the time, I realized that, from the standpoint of operational effectiveness and employment flexibility, the Air Force would be better off equipping the B-36 with four or five smaller weapons of, say, 2 megatons each. I suspected that the "requirement" for a 10-megaton weapon had more to do with the fact that the Navy's carrier-based aircraft would not be capable of delivering such a weapon than with considerations of operational needs, but I was told to "get with the program."

My job was not to develop the weapon itself but rather to determine how the bomber could safely deliver it. The proposed weapon was so powerful that the heat generated by its detonation could destroy the bomber that delivered it if provision was not made to ensure a safe escape distance between actual ground zero and the bomber.[1] People at Wright Field (then home to the Air Research and Development Command) had directed Allied Research Corporation to determine how much thermal energy a B-36 bomber could withstand while airborne. This is not a straightforward matter, since the answer depends on a host of variables, including the ambient air temperature, the flow of air over the aircraft's surfaces, and the rate at which the thermal energy is deposited on the aircraft's surfaces.

Using the estimates developed by Allied Research and the Atomic Energy Agency's projections of the yield of the hydrogen bomb, I devised an experiment, the purpose of which was to verify (or at least gain greater confidence in) the minimum safe escape distance for the B-36. We would fly an actual B-36 in an orbit placed just far enough from an atmospheric test of a hydrogen weapon so that the aircraft would receive no more than 80 percent of its limit load of thermal energy from the detonation.

The test took place at Enewetak Atoll in the South Pacific in the fall of 1952 and was, from the standpoint of the weapon's designers, a great success. From my standpoint and that of the crew assigned to fly the bomber, however, it was nearly a disaster. During the test, three things happened that, in combination, almost destroyed the aircraft: The weapon, which was itself experimental, had a yield almost double what was anticipated; the portion of the yield that constituted thermal energy was greater than expected; and the crew of the bomber was mistakenly at a position closer to the mushroom cloud than they were supposed to be. As a result, the bomber received a flux of infrared (IR)

---

[1]   Other products of the detonation, including gamma radiation and blast, had a shorter range and so were not of concern to us.

energy so powerful that it destroyed several panels on the side of the aircraft that faced the detonation. The crew managed to land the aircraft but it never flew again. Upon landing, they had some emotional words for the lieutenant colonel who had devised the experiment.

Developing the MB-1 Rocket

In the 1950s, defending the United States against a potential Soviet nuclear attack was a primary focus of U.S. defense planners. The threat of nuclear attack at that time stemmed solely from Soviet bombers: ICBMs were still in the early stages of development. In the United States, the Army was fielding SAMs capable of shooting down high-altitude bombers. The Air Force was responsible for maintaining interceptor aircraft on alert. Because guided air-to-air missiles had yet to be invented, the interceptors were armed solely with guns—and were judged as relatively ineffective.

In 1953, I argued that our interceptors could be substantially more effective if they were armed with nuclear-tipped air-to-air rockets. Because of the lethality of the warhead (gust loading would be the primary kill mechanism), these weapons could bring down a bomber without having to actually hit it. This would give Air Force air defense interceptors a much larger P_k, given an engagement.

Maj Gen Yates, the Chief of Research and Development on the Air Staff, took to this idea right from the start. In a few months I was assigned to the Air Force's Special Weapons Center at Kirtland AFB in New Mexico and placed in charge of initiating the program to develop and acquire such a missile. It would be called the MB-1 rocket, or the Genie.

One of the first priorities in determining the feasibility of the concept was to conduct a detailed analysis of the P_k that could be achieved. Maj Rex Mack, who worked for me, undertook this effort. This was no simple task. We needed to know what gust loads would be required to cause structural failure of the various elements of a bomber. Obviously, we also needed to know the character and intensity of these gusts as a function of distance and also as a function of the yield of the warhead.

The engineers at Wright Field were to provide answers to these questions and to render a report.

Within maybe three months, we had Wright Field's report in hand. It defined the "lethal volumes" for the wings; for the vertical and horizontal stabilizers; and for detonations in front of the aircraft, to the rear, below, and above. We used these volumes to determine an "equivalent" sphere—a sphere whose center is at the center of the bomber and whose volume is equal to the sum of all the volumes described earlier.

This "equivalent lethal volume" was a useful analytic construct. It would allow us to determine mathematically the probability that the warhead carried by our rocket would detonate within the sphere (the target) and thus to calculate the P_k for a given engagement.

Once the lethal volumes were revealed, in a document released by Wright Field, others examined this problem—in particular, Dr. Dike and Dr. Wood of the Sandia Corporation of Albuquerque, New Mexico, as well as a group headed by Dr. Milton Plesset and Dr. Lynn Gore of the physics department at RAND in Santa Monica, California. Both of these groups quickly reached the conclusion that the lethal volumes announced by Wright Field were seriously overstated—probably by a factor approaching five.

The group at RAND went a step beyond. They calculated the P_k by a method that I will label the *stochastic method*. They conducted a Monte Carlo simulation to determine the point in space at which the missile warhead *actually* detonated in repeated trials. The Monte Carlo had as inputs (1) the CEP of the point of detonation, at which the plane is perpendicular to the path of the missile, and (2) the linear error probable of the point along the path of the missile. For each trial they examined whether or not (yes or no) the detonation point was in one of the lethal volumes (or lobes).

After many trials they observed how many times "yes" and how many times "no." They could do all this because they were using a very capable computer—a rarity in those days. They found that the stochastic method yielded a much higher P_k for the same size lethal volumes than was obtained by the "math" method (equivalent sphere) that Major Mack had used.

The two analysts at Sandia Corporation arrived at the same conclusion: The stochastic method was the method to use, and the math method seriously understated the P_k. It turned out, somewhat providentially, that using the "smaller volumes" and the stochastic method yielded about the same P_k (around 0.80) that was obtained using the larger volumes by Wright Field and the math method (equivalent sphere). In our analysis, we had obtained the right number for the P_k, but not for the right reasons. We had two compensating errors. We promptly revised our analysis accordingly.

In addition to the yield of the warhead, the effectiveness of the system would depend on the speed and accuracy of the rocket delivering the warhead. The thrust of the rocket was already fixed: We were to use an existing rocket motor. The rocket's velocity, then, would be a function of the weight of the warhead. The accuracy of the fire control system was fixed as well: It was whatever the fire control system on the F-89J provided. This meant that the only parameter up for grabs was the yield (and, hence, the weight) of the warhead.

This presented a most interesting and critical trade-off: If you increased the yield, the warhead became larger and heavier. A heavier and larger warhead, in turn, reduced the velocity, both because of aerodynamic drag and because it weighed more ($F = ma$). A slower missile would take longer to reach its target and this increase in flight time would increase the "offset," the distance between the point where you thought the bomber would be at the time of detonation and where the bomber really was. We had to consider this offset, since, by direction, we were to assume that the bomber might be in a turn during the engagement, a tactic of zigzagging to make it harder to track and kill. So the time of flight was critical: The offset went as the time squared; that is,

$$s = \frac{1}{2}at^2,$$

where s is the offset, a equals acceleration, and t equals time.

When taking all these factors into account, we came up with the "finding" that there was indeed an optimum yield for the warhead, namely, 1.8 kt. We knew that meeting 1.8 kt required a warhead of a certain weight and diameter, given the warhead design technology of the day. By stating this diameter and weight, the engineers at Los Alamos Laboratory could determine the yield, and the engineers at Douglas Aircraft Company could calculate the velocity. By an iterative process, we determined the "optimum yield" and thus the optimum weight and diameter for the warhead.

We then defined the performance parameters (features) of the overall concept:

- warhead of 18.25 inches in diameter with an expected yield of 1.8 kt
- expected velocity of 3,200 ft/sec on average over the first five seconds
- a mil error (the aiming error) for the fire control system of 24, which was a given from the start.[2]

It turned out that the trade-off between yield (or weight) and velocity that we had determined prior to receiving the analyses by Sandia and RAND remained valid, even though we had been working with inflated lethal volumes. So, fortunately for us, the performance parameters we had specified remained valid even for the reduced volumes.

The MB-1 project had gained visibility at the Pentagon, and some colonels on the Air Staff felt compelled to issue a document about "requirements." Accordingly, they paid a visit to Kirtland. I showed them our analysis, underlining the point that there was an optimum yield—namely 1.8 kt. Any yields greater than 1.8 kt resulted in a lower P_k because the time of flight was longer.

The colonels were not impressed. They seemed to dismiss the whole analysis, stating it was "too technical." They announced that they had addressed this matter in a more "operational" manner. I replied that nothing could be more operational than to maximize the P_k within the technical constraints imposed, but to no avail. They

---

2   A *mil*, or *milradian*, is 1/1,000th of a radian and is the unit of measurement used for judging distances with an appropriately marked scope.

returned to Washington, and in due time, they issued a document that was intended to govern the further development of the MB-1 system. Nowadays we would call their document an operational requirements document (ORD). Their "ORD" stated that the "requirements" for the system were as follows:

- 2.5 kt yield
- 3,600 ft/sec velocity over the first five seconds
- 18 mil error for the fire control system.

I was totally dismayed when I received this document. The first problem, of course, was that their "requirements" could not be achieved: The "requirement" for a warhead with a yield of 2.5 kt was especially pernicious. If we held Los Alamos to that number, the warhead would be so large that the velocity of the rocket would be down to less than 3,000 ft/sec. The time of flight and the "offset" would be correspondingly larger, and the P_k would be greatly diminished.

I had already issued direction to Los Alamos and to Douglas that the warhead would be 18.25 inches in diameter. I was very reluctant to change this direction, since I knew this diameter warhead yielded the largest P_k within the constraints imposed. The colonels, when told repeatedly that the technology was not at hand to achieve their performance parameters, blithely announced that pushing forward the state of the art was, after all, the purpose of a development program. They failed to make the distinction between a "technology project" and a full-scale development program.

In light of this, I elected to ignore the newly stated "requirements." I was hoping that the storm would pass. The gambit to ignore their directions came to an end when, once again, the colonels appeared at Kirtland. They wanted to know if I had redirected the program to meet the "requirements" according to their document. I stated that I had not. I argued again that technology is technology: We could not possibly boost a warhead that provided 2.5 kt to a velocity of 3,600 ft/sec with the rocket motor at hand. They were defining performance features that simply could not be met. They told me to stop telling them what I couldn't do. "Just meet the requirements." As the meeting drew to a close, they demanded to know when I intended to issue new directions to the contractors. I replied that there would not be new directions as long as I was running the project, adding that, if they were determined to change the program to reflect their new parameters, they would have to get a new manager and have me fired. "We will do just that," they said. I had made a serious tactical error.

A short time later, the Chief of Staff of the Air Force directed the Air Force Scientific Advisory Board (SAB) to form a group to look into the "problems" attendant to the MB-1 project. This action was prompted by the colonels whom I had defied. By now, they realized that the case against me could not rest on the issue that I had refused to do their bidding. Rather, they would have to show that I was wrong. To this end, they secured the services of a "Dr. X" to critique my analysis. He secured a copy of our first cut and quickly determined what he called a fatal flaw in our analysis. We had (he believed) used the Wright Field volumes to calculate the P_k. Since these volumes were far too large, our analysis was seriously flawed. The P_k, he decided, was much smaller than we had calculated—around 0.50 or smaller.

When I found out that Dr. X had reviewed our first-cut analysis and not our more recent revision, I called him to tell him that when using RAND's stochastic method, the P_k was still something like 0.80, even when using the smaller lethal volumes. Without listening to what I had to say, he cut the conversation short. "Let us get right to the point," he said. "Do you still contend that a 1.8 kt warhead, a velocity of 3,200 ft/sec, and a 24 mil error yields a P_k greater than 0.50?"

"Yes!" I replied.

"That is all I need to know," he said. End of conversation. He would soon come to regret this bit of arrogance.

In due time, the SAB convened at RAND's offices in Santa Monica. Dr. X opened the proceedings with a briefing about his findings. It was too bad, he said, that I had used (in my calculations of the system's effectiveness) the lethal volumes as determined by the engineers at Wright Field. He acknowledged that I had no other choice, since I was directed to do so. But Wright Field's estimates were flawed and so, unavoidably, was my analysis. Gratuitously, he said that a mistake of this sort was to be expected from an analysis led "by a colonel unschooled in operations research supported by four lieutenants with Friden calculators"—a mechanical marvel of the day.

My opportunity to rebut the statements by Dr. X came in the afternoon. In my presentation, I led with a chart that stated in bold letters, "The lethal volumes used in the analysis we now stand behind assume smaller volumes than those used by Dr. X. But, when we use a 'stochastic' approach, we find that, even with the smaller volumes, the P_k is more than 0.75." Then came the bombshell: "The analysis we now stand behind was done by analysts from RAND and from Sandia."

I had changed the terms of the argument. The argument now was not about lethal volumes. Rather, the argument was between using a math method (the equivalent sphere) versus using a stochastic method. The two analysts from Sandia and the two from RAND (all four were in the audience) supported me on this point. Dr. X was thoroughly rattled and ill prepared to join this argument.

I then pressed on and engaged in a gambit for which I was criticized later—even by some of my supporters. "Dr. X," I began, "one of the criticisms you levied in your critique was that our analysis was mathematically inelegant. To bolster your argument that our analysis was seriously flawed, you made the statement, which is copied verbatim on the next chart." That chart stated, "'They used as the criteria for stand-off (safe escape) that the median dose of radiation received by the pilot was not to exceed 25 rems. A better criterion would have been that the pilot never receives more than 35 rems.'"[3]

"Since you issued that critique," I said, "you have had the opportunity to learn much more about this matter. Would you now like to change or withdraw that statement about the criterion for safe escape?" Dr. X answered, not unexpectedly, "No!"

I then observed that I had asked my lieutenants which was the more demanding criterion for "stand off": that the median dose the pilots would receive would not be more than 25 rems, or that the pilot

---

[3] *Safe escape distance* refers to the distance required between the interceptor and the exploding warhead to ensure that the pilot of the interceptor would not receive a harmful dose of radiation as a result of using the MB-1 system. The term *rem*, coined from *roetgen equivalent man*, is the unit used to measure radiation absorbed by human tissue.

would "never" incur more than 35. They said they did not know how to deal with "never" in a bivariate normal distribution. They could calculate two or three standard deviations from the mean, but not "never." At this point, Dr. X saw the trap. But it was too late: He had seen no reason to change the statement when it was on the screen a minute before. Dr. X had carelessly used the (incalculable) term *never*. I then admonished that anyone sloppy enough to use the term *never* in this way had no business accusing me or my lieutenants of mathematical inelegance. "You owe those four lieutenants an apology," I concluded.

At this point, Dr. X appealed to Dr. Puckett, the head of the SAB panel, saying that he refused to be harassed in this way. In response, I declared that we could put a stop to the harassment promptly: "The way to bring this whole affair to a stop is for you to cease and desist and agree that my analysis is elegant and correct and that your analysis is fatally flawed."

Dr. Puckett intervened. But we were home free. The panel reported that the directions that I had given to Los Alamos and Douglas had been correct, adding that I had acted properly in not adhering to the "requirements" as set forth in the "ORD." The three colonels from the Air Staff were in the audience, but they remained mute. The program was now free to proceed without interference.

Shortly after the SAB meeting, I was selected to attend the Air War College, and Col Bill Black of the Development Directorate at Kirtland took over the responsibility of running the program.

From the standpoint of research and development (R&D) and acquisition, the program was a success. The MB-1 rocket was developed and produced (1) ahead of schedule, (2) under cost, and (3) on target with regard to expected performance. On the other hand, it was a disappointment operationally. Unbeknownst to me at the beginning, the technology for providing air-to-air missiles with terminal guidance was being pursued at the Naval Weapons Center under the cloak of a very secret project. This technology matured. The AIM-9 Sidewinder missile was developed and acquired, and the MB-1 rocket was shoved aside. It would be irrational to use a weapon that involves the detonation of an atomic warhead when guided air-to-air missiles with high-explosive warheads and a comparable P_k were available.

Something like 3,000 MB-1 rockets were produced. Some Genies (disarmed, of course) are still around but in obscure corners of museums.

Other Observations

As a result of my arguments with the Air Staff over the key performance parameters of the MB-1, I came to have a distinct dislike for the word *requirements*. There is great opportunity for folly when you approach the problem of labeling the expected performance features of the systems attendant to some concept as *operational requirements*. Too often, the term takes on a life of its own: "This is what the operator needs. Quit arguing, period." This invited the idea that the concept

---

Box 4.1
Lessons from Running the MB-1 Project

It Pays to Be Lucky. I had used the wrong values for the lethal volumes in my first cut and calculated a high P_k. When using the new and smaller volumes, the stochastic approach bailed me out. The P_k was still calculated to be high. I originally had approximately the right answer but not entirely for the right reasons.

Seek Outside Help from Experts. Both RAND and Sandia put a great deal of effort into determining the critical parameter—the P_k. Their expertise ensured that the program was on solid footing analytically. It also helped later on to fend off the assaults of the "requirements boys" on the Air Staff. Of course, just bringing in outside experts is no guarantee of success, even if they really are experts. The key is in knowing what factors are most important to the performance of the overall system and zeroing in on the possible trade-offs among them.

Do Not Bow to Bureaucracy. If you are confident in the integrity of your analysis (and you should be), be prepared to go broke on your own stupidity, not that of some faceless person as reflected in some document. Acceding to the demands of the Air Staff would unavoidably have induced great overruns in both time and money in the program.

A Good Offense Is Better Than a Good Defense. It is more fruitful to attack the critique of your analysis than to try to prove that your analysis is without error.

---

is a failure if it does not meet the performance factors that have been named "operational requirements."

The problem stems from giving the name *requirements* to what are really *performance specifications*. The word *requirement* suggests no flexibility. A better term would be *expected performance characteristics*. Nevertheless, the problem inherent in specifying expected performance as operational requirements persists. Indeed, it plays a central role in DoD's formal process for defining new development programs. The next section illustrates another case in which misuse of the word *requirement* caused undue turmoil in an important acquisition program.

The Short-Range Attack Missile Affair

Another program that incurred unneeded turmoil as a result of the misuse of the term *requirement* was the program to develop a short-range attack missile (SRAM) for the B-52 strategic bomber. The program began in the mid-1960s. The concept was to equip the B-52 with a nuclear-armed missile to attack a Soviet SA-3 air-defense site with impunity. The SRAM would be fired at the site from a distance beyond that at which the SA-3's radar at the site could detect and track the B-52.

The calculation for the "required" range was straightforward. At 23 nmi, an aircraft flying at 400 feet above ground level cannot be detected and tracked by a radar on a 90-foot tower with a grazing angle of 6 degrees to the horizon. At a lesser range, the radar could detect and track the aircraft. So the 23 nmi range was indeed an operational requirement. If the range of the missile was less than 23 nmi, the SA-3 battery could not be attacked with impunity, and the concept would be null and void.

A development program was commenced. By about the late 1960s, the SRAM was approaching initial operational capability (IOC). I was in AFSA at the time. A colonel from another office brought in a report for my coordination. It was an annual report to Congress as to the status of the acquisition programs the Air Force was conducting, system by system. For each system, there was an item called "Requirement." To my surprise, the "requirement" listed in this document for the SRAM was that the missile have a maximum range of 37 nmi. I had been involved in framing the concept for the SRAM some years before and remembered that calculations determined the maximum range to be 23 nmi. After some investigation, we learned that the switch to 37 nmi came from a calculation by an engineer at Boeing, the prime contractor for the program, who determined that the system they were building would have "thrust" and "drag" such that it should be expected to fly 37 nmi. I pointed out that this longer-range figure was a statement about expected performance, and, harkening back to my earlier experiences, I added that there was some danger in putting it down as an "operational requirement." The colonel agreed in good faith to change the draft he was coordinating to state that the requirement was 23 nmi, but his superior, a general running the directorate for "requirements," overruled him. "Why put down 23 nmi when we are on contract for 37?" he reasoned. They did not inform me of this decision.

A year went by. A new annual report went to Congress, one that I did not see before it was dispatched. In this report, the Air Force listed next to the entry for the SRAM that the range "requirement" was now 33 nmi. In the year since the previous report, the engineers at Boeing had encountered some unexpected problems, and the thrust they were able to achieve went down, while the drag went up. The system now under development did not now meet the previously stated requirement of 37 nmi. A staffer for Rep. Samuel Stratton (D-NY) spotted the change from 37 to 33 and informed the congressman: Congressman Stratton went public, stating that the missile cannot meet the required range and should be cancelled. The matter was referred to the U.S. General Accounting Office (GAO) for investigation.

By now, the whole affair gained the attention of the Chief of Staff of the Air Force. He gave me the job of stopping the GAO report. I was not successful; the GAO stated that 37 nmi (the required range) could not be attained. In fact, according to the GAO, even the 33 nmi range was in doubt. The office recommended the program be cancelled. Being accountants, they knew the difference between two numbers but not the relevance of either. In due time, Congressman Stratton held hearings before a rump session of the House Armed Services Committee. At those hearings I explained that the original calculation of the required range (23 nmi) was based on such operational matters as the need to be able to attack Soviet SAMs with impunity. I stated that, if the expected range became as low as 25 nmi, the Air Force on its own would cancel the program.

Congressman Stratton replied that my presentation helped to clarify the distinction between a genuine "operational requirement" and "expected performance." In light of this distinction, he expressed bewilderment at the fact that the Air Force had ever stated that the requirement was 37 nmi and asked me how such a thing could happen. In response, I blurted out, "Because we have our share of people who sometimes do not think straight." To this, the congressman replied, "If that be the case, then I expect you to take remedial action." "Done," I said.

Following the hearing, I reported to the chief that Congressman Stratton no longer considered the SRAM an issue. However, I noted that the congressman enjoined us to take action so that colonels who do not always think straight have no part in preparing documents that are sent to Congress. The chief dutifully gave a brief statement at the next staff meeting about the danger of confusing operational requirements with expected performance, and the colonel was transferred out of the Pentagon. Even so, the practice of blurring this distinction continues and is still alive and dangerous today.

The Minuteman Missile

In the late 1950s, the Air Force was going about the business of organizing, staffing, and equipping units (squadrons) with ICBMs. These units were to be equipped with either the Atlas or Titan missile. Gen Bernard Schriever was the head of the organization to develop, manufacture, and deploy these missiles. The concept was to fabricate segments of each missile in a large plant, transport these segments to the site, and then assemble the missile in the silo—a Herculean endeavor.

General Schriever, a man of uncommon insight and vision, tasked his staff to define and evaluate a better concept:

1. Use a solid propellant.
2. Assemble the missile at the plant.
3. Transport the "integral" missile by air transport to a base near the silo.
4. Transport the integral missile to the silo over country roads in a vehicle known as the transporter-erector.
5. Insert the missile in the silo by the same vehicle.

The missile was to have three stages and a single RV.

Today, this concept seems like a piece of cake—and it is. But when the concept first emerged, there was some doubt as to whether it all hung together. There were questions about whether or not it was possible to build a solid-propellant missile that was small enough to be transported over country roads but at the same time would produce enough thrust to propel a certain weight RV some 5,500 nmi—a heady proposition.

After some tests of solid propellant motors and several designs of the transporter-erector, General Schriever decided to "move." Moving meant going to the Pentagon and gaining approval to implement the concept. His immediate targets were the Chief of Staff and the Secretary of the Air Force. General Schriever did his homework regarding the approach to seeking approval to implement his concept. He found that General White, the Chief of Staff, had (rightly) placed uncommon trust in the recommendations of General Gerhart, his Deputy for Plans and Programs. In turn, he found that General Gerhart relied heavily on the advice of a certain Colonel Kent. Bingo. General Schriever knew me and, in due time, he dispatched a Colonel Lulejian to brief me and no one else in the Pentagon on the new concept. Later in the day, I gained an appointment with General Gerhart. After the briefing by Colonel Lulejian, General Gerhart said he supported the concept. So Colonel Lulejian reported to General Schriever: "Mission accomplished."

A week or so later, General Schriever took the Air Staff by storm. By late afternoon he had an audience with the Secretary of the Air Force, James Douglas, known as the Judge. (He was at one time a judge of some renown on a federal court.) The Judge listened intently, asked probing questions having to do with technical feasibility, and then, after more than two hours, announced, "The court rules that you [General Schriever] are to proceed to implement this concept with all due and prudent haste. I will square it with Dr. Quarles [the Deputy Secretary of Defense] and undertake to gain the appropriate allocation of resources." From a "dead start" to that galactic announcement—all in one day. The rest is history. The Minuteman played a key role as a means of implementing our strategy of deterrence in the Cold War.

The example shows what can be done, and done correctly, by decisive advocates dealing with decisive deciders who share a common goal and a common trust. No effort or time was wasted in developing a mission needs statement (MNS) or an ORD, an analysis of alternatives (AoA), or other time-consuming documents. Of course, not every proposal can be handled with such dispatch. But the Minuteman case shows how key aspects of today's burdensome "requirements process" could (and should) be streamlined:

- If the Secretary of Defense were to publish a short list of the highest priority operational needs, development efforts and leadership attention could be focused on these, and debates about the relevance of certain proposals could be short-circuited.
- A more-deliberate and -coherent process of concept development in the services could yield proposals that are more mature when they reach decisionmakers.
- The secretary needs greater flexibility in allocating resources so that the services have strong incentives to promote new and promising concepts and gain approval to implement selected concepts quickly.

Responding to a Possible Soviet Nationwide Antiballistic Missile Deployment

In the mid-1960s, the intelligence community observed that the Soviets were constructing what was believed to be an ABM complex near Tallinn, the capital of Estonia. There was speculation that this complex "might be" the precursor to a nationwide ABM system.[4] After attempts to gain collaborative evidence, the statements about this site were along the lines that it could be, probably was, or surely was a precursor to the Soviets deploying a nationwide ABM system.

Dr. Brown, Director of DDR&E, decided that the United States should take some form of action in response to this possible emerging threat. He fashioned a one-page letter for the signature of Secretary McNamara, directing him (Dr. Brown) to develop "options" we should pursue. The secretary added his own statement to the letter, a paragraph to the effect that he himself was not convinced that the Soviets had the capability or the intent to deploy a nationwide system that would seriously degrade our retaliatory attack. But, even so, we should now initiate efforts to counter such a threat since the downside of waiting could be great.

Dr. Brown, upon receipt of this letter, gave me the responsibility of spearheading the effort to develop and define concepts for us to pursue. I interacted with several people: VADM Levering Smith of the program office for Polaris; General McGraw, the program manager for the Minuteman; and selected people at the Aerospace Corporation, RAND, and TRW Inc., including Dr. Ernie Krause.

In less than a month, I fashioned a letter from Dr. Brown to Secretary McNamara. The letter recommended that we pursue the following efforts:

1. Develop and field a new and larger missile for our existing submarines so that we had more throw-weight and could deploy more RVs and decoys.
2. Develop and field a new front end for this new and larger missile. The new front end would have multiple warheads.
3. Develop and field a new front end for the Minuteman missile. The concept was to have a post-boost vehicle (PBV) that could deploy several RVs and decoys, targeted at different DGZs, or aimpoints.

---

[4] This is the same system described in "Defending the Planners of the SIOP," pp. 30–37.

4. Upgrade the third stage of the Minuteman missile to provide more throw-weight.

5. Initiate a "technology project" designed to conceive and nurture new concepts for countering Soviet ABM systems. This project was eventually known as Advanced Ballistic Reentry Systems and centered on developing decoys that could withstand reentry and, at the same time, did not weigh much.

Dr. Brown promptly dispatched this letter to McNamara. In less than a week, the secretary sent the letter back with "OK RMC" opposite each option. And so the Poseidon missile and the MIRV-equipped Minuteman (Minuteman III) became a reality.

A larger missile with multiple RVs for the submarines and a MIRV-equipped Minuteman were concepts that were being considered independently of the mandate to counter any Soviet ABM system. The destructive power would be increased by deploying more than one RV. The directive from Secretary McNamara to Dr. Brown was the trigger for purposeful action.

In the presence of the "OK RMC" for each option, Dr. Hitch directed the PPBS to do the programming and allocate resources to the appropriate line items in the budgets. The initials were all that he (and we) needed to proceed. Again, as with the development of Minuteman in its first incarnation, there was no need for documents, such as an MNS, ORD, or AoA—steps that often waste time and effort. The process of responding to the emerging threat had timelines that were so short as to be dazzling by today's standards. Progress was rapid because the principals involved were both knowledgeable and decisive and took purposeful actions, generally in the form of letters or directives.

Happily, this episode ended well. Secretary McNamara was right: The Soviets never deployed a nationwide ABM system worthy of the name. But we had hedged against the possibility of a "breakout" deployment of Soviet ABM systems and, at the same time, increased the destructive power of our ICBM force by deploying multiple RVs that were independently targetable on one booster.

The Minuteman III

From the very beginning of the program, the number *three* was bandied about as the number of RVs for the front end of the new Minuteman III. But Dr. Brown felt that we needed better insight into this matter. To this end, he tasked the Minuteman SPO to conduct an analysis to provide better insight into the matter of defining the front end of the Minuteman III. In due time, they came back with an analysis that was more complicated and extensive than it was insightful. Dr. Brown sent them back to the drawing board. Weeks later, they again appeared in his office. Again he rejected their analysis—not so much because it gave the wrong answer, but because it was neither insightful nor compelling.

Upon their departure, he directed me to go out to Norton AFB, California, the home of the Minuteman SPO, and stay there until they produced an analysis that he could use to defend the number of RVs per missile on the Minuteman III—whatever the number was. I pointed out that there were two separable issues: (1) creating an analysis that would meet Dr. Brown's standards and (2) having the SPO produce that analysis. I blithely announced I could provide the analysis, but I was doubtful of my ability to get the SPO to do an analysis that passed his muster.

"All right," he replied. "What is it?"

"Sir, it is a matter of elimination. The candidates range from zero to five. By inspection we can eliminate zero."

He failed to see the humor in this.

"It can't be one. We already have that covered. It can't be four or five, as calculations by our engineers show that, if the missile is to boost a PBV with four RVs 5,500 nmi, the weight of each RV is so constrained and the nuclear warhead is so small that the yield goes down the drain. That leaves us with two or three."

To choose between two or three, we must decide on the value of the "scaling factor." The scaling factor of one-half controls the trade-off between number of RVs and the yield of each when attacking an area containing "soft" targets. The scaling factor normally used was two-thirds, but I had shown previously that it was more like one-half. The derivation of the scaling factor for nuclear weapons is summarized in

"The Trade-Offs Between Numbers, Yield, and CEP in Hard-Target Kill," pp. 226–229.

When using the scaling factor of one-half, two RVs or three RVs were about a tie as far as destructive power was concerned: That is, a Minuteman missile with two RVs had about the same destructive power as a missile armed with three lower-yield RVs. But the three RVs were obviously preferred in terms of countering an ABM system. First, more RVs meant that there would be more objects for the Soviet ABM system to counter. And, second, if in the future it was determined that decoys would need to be deployed, the decoys appropriate for simulating the smaller warhead would presumably be lighter than those called for by a two-RV design, so there could be more of them for the space or weight available. Thus, three was better. Dr. Brown concurred.

Later, Dr. Brown revisited the matter of the Minuteman III. He wanted insight as to the number of missiles to deploy and the number of RVs to deploy on each missile. Since he, in effect, had asked two questions, this indicated that the number of RVs per missile was not entirely settled.

After several false starts, I arrived at the following construct: The requirement is to have some number of RVs (x) survive a Soviet attack. The Soviet attack is defined by how many Soviet RVs are assigned to the attack of the Minuteman force and the probability of an RV killing a Minuteman silo once assigned.

Empirically, from engineering data, I took that whatever a one-RV missile may cost, a four-RV missile costs twice as much, and a nine-RV missile three times as much, a scaling by the square root. If you deploy very large missiles, the cost per RV deployed goes down, but the number of aimpoints also goes down, and the expected fraction of RVs surviving the attack also goes down. So there must be an optimum number of RVs per silo (missile)—optimum in the sense of having a stated number of RVs surviving a stated Soviet attack and doing so at least cost.

After several futile attempts, I solved the problem. There was an "optimum" survival percentage, and that number was, interestingly enough, $\frac{1}{e}$ (that is, one over the Napierian e, or 0.37). So, the number of RVs to deploy was the "requirement" times e, or times 2.718. On the other hand, the number of silos to deploy was such that the quantity $(1 - P_k)^n = 0.37$, where n is the number of Soviet RVs assigned to the attack on the Minuteman divided by the number of Minuteman silos deployed.

Thus, if the requirement is that 600 RVs are to survive, then we should deploy

$$600 \times 2.718 = 1{,}631 \text{ RVs.}$$

If the threat is 800 Soviet RVs with a P_k of 0.5 each, then deploy 552 silos, since

$$\frac{800}{552} = 1.45$$

and

$$0.5^{1.45} = 0.37 = \tfrac{1}{e}.$$

I showed this math to Dr. Brown, and he announced the analysis was clever—maybe too clever. We would get different answers if we changed the requirement or the threat. That is so. But reasonable cuts at these two parameters provided a rationale for what we were planning, namely 550 silos and 1,650 RVs.

I recite this example to underline the idea that, if the analyst works hard enough, and long enough, and clearly enough, he or she will eventually arrive at a simple analytical solution to a complex problem.

Defining the Deployment of the Minuteman III

As stated in the prior section, the problem, in the barest of terms, is to deploy enough RVs on enough missiles, M_{dep}, so that a required number of RVs, Q, survive a stated Soviet attack on the Minuteman silos deployed—and do so at least cost.

With regard to cost, we know empirically that, if a single-RV missile costs one unit, a four-RV missile costs two units. In general, the unit cost per missile equals the square root of n, where n represents the number of RVs per missile. One can readily see that, for a stated value of Q and a stated value for the size of the Soviet attack, there will be an optimum number of RVs per missile that will minimize the cost of the deployment. More RVs per missile will reduce the cost per RV deployed, but more RVs must be deployed, since the P_s, the percentage of Minuteman missiles surviving the attack, is reduced. With bigger missiles, there are fewer silos to attack and thus more Soviet RVs per silo.

After several futile attempts to derive the optimum number of RVs per missile (using partial derivatives), I hit upon the following approach:

1. The cost of the deployment, z, equals the number of missiles deployed, M_{dep}, times the cost per missile, c_M,

$$z = M_{dep}\, c_M .$$

2. Postulate a value for P_s, the fraction of Minuteman missiles surviving the attack.

3. The number of RVs deployed, R_{dep}, is then

$$R_{dep} = \frac{Q}{P_s} .$$

4. The number of missiles deployed, M_{dep}, must equal

$$M_{dep} = \frac{S\left[\ln\left(1 - P_k\right)\right]}{\left(\ln P_s\right)},$$

where S is the number of Soviet RVs and P_k is the single-shot probability of kill of each RV against a silo.[5]

5. Then, the RVs per missile is simply

---

[5]   This equation stems from the fact that, for a silo, $P_s = \left(1 - P_k\right)^{\frac{S}{M_{dep}}}$.

$$\frac{R_{dep}}{M_{dep}},$$

and the cost per missile is then

$$c_M = \sqrt{\frac{R_{dep}}{M_{dep}}}.$$

Putting all the above together,

$$z = M_{dep}\sqrt{\frac{R_{dep}}{M_{dep}}}.$$

Thus,

$$z = \sqrt{\frac{Q \times -\ln\left(1 - P_k\right) \times S}{P_s \times -\ln P_s}}.$$

All parameters in the equation above are fixed except for the P_s. If we are to minimize the cost, then, we must maximize the quantity $P_s \times -\ln P_s$. We know from math that x times the $-\ln x$ is a maximum when x equals $1/e$. The beauty of this solution is that the optimum P_s $(1/e)$ is constant. It does not change if Q is changed or if the Soviet threat is increased or decreased. On the other hand, the optimum number of RVs per missile will be different depending on the value of Q and the severity of the Soviet threat.

My first attempt at solving the problem had been to find the "optimum number of RVs per missile" using LaGrange multipliers. I failed to arrive at a satisfactory answer using this approach and finally realized that the factor that remained constant was the optimum P_s.

A few years later, I used the insight that there was an optimum P_s to good effect. At a gathering of analysts sponsored by the Military

Operations Research Society, an analyst with connections to the Navy presented a briefing that showed, among other things, that only 40 percent of the Minuteman III missiles would likely survive a Soviet attack on U.S. ICBM silos. My comment was that I realized we were a little off from the optimum—the optimum being that 0.37 of the Minuteman missiles should survive. But if we wait a bit and the Soviet threat increases, the P_s of Minuteman missiles will be right on target. The analyst who had given the briefing had no idea what I was talking about and declared that my idea to plan for 0.37 to survive was "preposterous." I offered to send a copy of the derivation to anyone who asked, but somewhat to my chagrin, there were few takers. This solution approach, which is elegant in its simplicity, was appropriate for the types of capabilities and numbers of missiles and RVs at the time of the analysis. However, the problem is solved without considering the integer requirement for the number of RVs per missile and does not constrain the solution to a reasonable number of RVs per missile. The optimal solution for the integer case or when constrained in terms of RVs per missile could differ significantly under some sets of parameter values.

Penetrating Soviet Air Defenses: The Argument for Decoys

There was continual ferment about the ability of our bombers to penetrate the Soviet air defenses as they proceeded to their appointed DGZs, according to the SIOP. Some argued for the concept of substituting decoys for weapons to dilute the "kill potential" of Soviet SAMs and Soviet airborne interceptors protecting these targets. For example, if there were one escort decoy for each bomber and if the Soviets were unable to distinguish between a "real" bomber and a decoy, the Soviet kill potential per bomber would be halved, and thus the fraction surviving of bombers (FSB) would be correspondingly increased.

Accordingly, we embarked on an analysis to gain insight as to whether employing decoys was a valid concept. Our general construct was as follows:

$$FSB = e^{-\frac{KP}{B+D}},$$

where KP is the kill potential the bombers will incur from both Soviet SAMs and Soviet interceptors while traversing a particular zone, B is the number of bombers, and D is the number of decoys. We had developed a simulator that captured the interactions of the Soviet SAMs and Soviet interceptors with the Blue bombers as they traversed each zone. The output of the simulator was the FSB, zone by zone.

Once we knew the FSB from the simulator, we could calculate the kill potential for each zone:

$$KP_{Sov} = \left(-\ln FSB\right)\left(B + D\right).$$

Now that we knew the Soviet kill potential for each zone, we could use the aforementioned equation to calculate the FSB for stated numbers of bombers and stated numbers of decoys. The effect of the decoys was to spread the kill potential presented by the Soviet SAMs and interceptors over a larger set $(B + D)$.

We presented a chart that revealed the outcome for five cases—the outcome being the number of "weapons on target" (see Table 4.1).

The illustration in the table note applies generally to all the cases. In cases A and B, the substitution rate is three to one. In cases C and D, the substitution rate is six to one.[6]

Table 4.1 led to some insights:

1.   The number of weapons on target (weapons to DGZ) is a maximum when one or two weapon spaces per bomber are allocated to decoys. This holds whether the substitution rate is three to one or six to one.

---

[6]   During the preparation of this book, one of my colleagues at RAND observed an error in the logic of our analysis. Specifically, when we did the analysis, we assumed that the FSB was the same for every zone. However, as the number of surviving bombers decreased with each successive zone, the FSB should decrease. This error meant that our calculations overstated the number of bombers surviving but actually understated the value of decoys.

Table 4.1
Quantifying the Value of Decoys Against a Layered Air Defense

Subst. Rate	Case	Weapons (EQ) (no.)	Decoys (EQ) (no.)	Decoys (Flying) (no.)	FSB (1 zone)	FSB (6 zones)	Weapons (to DGZ) (no.)
N/A	Base	80	0	0	0.861	0.407	32.5
3:1	A	70	30	5	0.905	0.549	38.4
3:1	B	60	60	10	0.928	0.638	38.3
6:1	C	70	60	10	0.928	0.638	44.6
6:1	D	60	120	20	0.951	0.741	44.4

NOTES: In each case, ten bombers seek to penetrate six defense zones. Each bomber has a capacity of eight weapons. The Soviet kill potential in each zone is 1.5; we got this number from the simulator.

The FSB of a particular zone is given by

$$FSB = e^{-\frac{1.5}{B+D}}.$$

In the base case, there are no decoys; thus, 80 weapons are carried.

In case A, however, there are 30 decoys: one weapon space per bomber is allocated to decoys at a substitution rate of three to one: 10 × 3 = 30 decoys.

Each decoy covers only one zone. Therefore, the 30 total decoys provide only five decoys (flying) at a time in each of six zones. The FSB for one zone, then, for case A is 0.905, and the FSB for six zones is 0.549.

The number of weapons on target (the DGZ) is 38.4 (0.549 × 70 = 38.4).

2. An increase of more than 35 percent in operational capability is gained at a substitution rate of six to one—44.6 weapons on target versus 32.5.

Preliminary analysis by engineers indicated that the substitution rate would probably turn out to be around six to one. That is, the engineers stated that they could produce a decoy that had the legs to traverse a "zone" as an escort decoy and that the bomber would have to give up only one weapon space to carry six such decoys.

I thought the analysis made a compelling case for decoys. But an official in DDR&E thought otherwise. He was of the opinion that he could not recommend that we proceed with a program to develop and acquire escort decoys based on such a simple analysis. "For one thing," he said, "I need to know the total number of decoys we intend to buy."

I replied, "I plead guilty to the charge that the construct is simple—once you gain FSB from the simulator and then calculate the Soviet KP. At the same time, I will argue that the construct we have used provides ample and correct insights on the issue at hand—namely, the decision to begin a program to develop and acquire decoys. The total number of decoys to be acquired and employed depends on factors we do not presently know, namely, an engineering factor, having to do with the trade between decoys and weapons, which we will find out during the development program. If it is six to one, then we should plan to field about six to twelve decoys per bomber; if three to one, then about three to six decoys per bomber. How many we finally acquire is a matter of how long we continue the production line. We can and will make that decision later."

I concluded by saying, "I do not claim to know with any precision how many bombers will penetrate to target, without decoys or with decoys. I do claim to know at this time that the optimum number of decoys per bomber is certainly not zero."

The DDR&E official was adamant. He was not about to give a go-ahead to a program to develop and acquire decoys based on our presentation. In fact, he became quite abusive about the lack of depth of our analysis.

Then in a week or so, fate intervened. For a number of reasons, the DDR&E official vacated his position. Shortly after he left we briefed Dr. John Foster, head of DDR&E. When we had finished he exclaimed, "This is elegant. We will undertake a program to develop and acquire decoys."

One lesson here is that the virtue of a particular analysis is sometimes in the eyes of the beholder. Needless to say, I agree with the assessment by Dr. Foster. Just because the analysis is simple does not mean it is not adequate to gain insight into the issue at hand. The decoy was developed and acquired in large numbers, and the Quail (its name) became an integral part of our portfolio of penetration aids.

This affair also illustrates how to use the output of a simulator at the engagement level as the central input in a more-encompassing analysis. We gained knowledge of the kill potential of the Soviet SAMs and Soviet interceptors by observing the FSB after penetrating a defined zone in the defense. The simulator was hard to run and was "noisy"— that is, being a stochastic model, it yielded results that varied significantly from run to run. From the simulator we gained knowledge of the kill potential per zone. Armed with this input, we proceeded as described.

Finally, this case shows that even sound analyses will sometimes be rejected by mulish decisionmakers. When the analyst encounters such a roadblock, he or she sometimes has the option of finding other audiences for the work. In this case, I lucked out, and the roadblock was removed for me.

Keeping Bombers in the Triad: The Mix Is the Thing

In the late 1960s, I was the Military Assistant to Dr. Harold Brown, the Director of Defense Research and Engineering in OSD. During this period, a DPM was prepared by Dr. Enthoven every year. The DPM, which set policy for nuclear forces, stated what types of forces we would deploy and in what numbers.[7] It addressed the forces we had already deployed and also those we should plan to develop and deploy.

For some time, Dr. Enthoven had argued that the triad should become a diad. He proposed in the DPM to eliminate manned bombers, retaining ICBMs and SLBMs as the two legs. He had an analysis, and according to his calculations, the SRAMs carried by B-52 bombers cost 1.5 times as much to deploy as the RVs on Poseidon submarines. He argued that, for an equal cost, the United States could deploy 1,200 RVs on Poseidon submarines, or 800 SRAMs on manned bombers.

He thought it self-evident that 1,200 RVs on Poseidons would be the better choice. On a per-weapon basis, the two different weapon systems had about the same potential to destroy enemy targets, and

---

[7] For a description of the DPM and its role in defense planning, see "Helping with DPMs," pp. 54–56.

the prospect of surviving a Soviet attack favored the RV on a stealthy submarine at sea over an SRAM weapon on a B-52 bomber on quick alert.

I might have challenged Dr. Enthoven's cost estimates. But the ensuing debate could have become very convoluted because it was so unclear just what costs should be included, especially with respect to the infrastructure for both weapon systems. Moreover, Dr. Enthoven had an advantage over me when it came to cost estimates. He was expert in cost analysis. He worked for the DoD Office of the Comptroller, which presumably knew all about costs.[8] So I let his cost estimate go unchallenged and argued instead that a mix of RVs and SRAMs might be better for the United States even if SRAMs cost more than Polaris RVs. Dr. Enthoven presented a table similar to Table 4.2.

He defined one unit of cost as the cost of 100 RVs. Because he had estimated that SRAMs were 1.5 times as costly as RVs, 18 units of cost were required for SRAMs to do the same job as 12 units of cost in RVs. He then calculated how many units of cost would be necessary against defended targets, assuming that, in the presence of defenses, it would take twice as many RVs or SRAMs to destroy the same number of targets. Of course, multiplying by two opened the gap between RVs and SRAMs even more; that gap went from six units of cost in the undefended case to 12 units of cost in the defended case. I felt that the analysis involving defenses missed the point about the value of a "mix."

Table 4.2
Cost to Attack 1,200 Targets

Weapon System	The Soviets	
	Have No Defense	Have Defense
Polaris RVs	12 units of cost	24 units of cost
B-52–launched SRAMs	18 units of cost	36 units of cost

---

[8] Dr. Enthoven had a doctorate in economics from MIT and had previously worked as an economist at RAND and in the Comptroller's Office of DoD. Indeed, his Systems Analysis Office had been part of the Office of the Comptroller before separating in 1965.

In preference to Dr. Enthoven's highly simplified approach, which was built entirely around cost, I proposed a more-sophisticated approach. I proposed that we analyze how many targets of the 1,200 would be destroyed under different strategies on both sides, still assuming that SRAMs were 1.5 times as expensive as RVs and further assuming that the Soviets would have to pay the same cost in defenses to defend a target either against an RV or a SRAM. However, the Soviet Union would have to decide whether to deploy interceptors designed to defeat RVs or interceptors to defeat SRAMs; the same interceptor could not do both jobs.

This more-sophisticated approach turned the tables on the analysis by Dr. Enthoven. He had introduced the concept of a nationwide Soviet defense, thinking it would make his argument more persuasive. But he had not reflected that the Soviets would have to build very different systems to defend against ballistic missiles (RVs) as opposed to rockets (SRAMs) delivered by bombers. Neither had he considered the effects of different strategic choices on both sides. In other words, he had opened the issue of Soviet defenses without thinking it through.

I maintained that the United States would have three options—all of equal cost—in this case, 1,200 units of cost:

- Option A: United States deploys 1,200 Polaris RVs.
- Option B: United States deploys 800 B-52 SRAMs.
- Option C: United States deploys 480 RVs and 480 SRAMs.

The Soviet Union would have the following two options:

- Option X: Soviet Union defends 1,200 targets with 1,200 interceptors against RVs.
- Option Y: Soviet Union defends 1,200 targets with 1,200 interceptors against SRAMs.

The outcomes for various combinations of strategies are displayed in Table 4.3, which shows targets destroyed for three options by the United States versus two options by the Soviets.

Table 4.3
Targets in the USSR Destroyed Under Different Combinations of Forces

| | Soviet Union (defense) | |
United States (offense)	Option X 1,200 RV interceptors	Option Y 1,200 SRAM interceptors
Option A 1,200 RVs	600	1,200
Option B 800 SRAMs	800	400
Option C 480 RVs 480 SRAMs	720	720

Case outcomes:

- **U.S. option A vs. Soviet option X:** 600 targets are destroyed. The United States sends 1,200 RVs against 600 targets that are defended by RV interceptors. In the presence of these Soviet defenses, it takes two RVs to destroy each defended target.

- **U.S. option B vs. Soviet option X:** 800 targets are destroyed. The United States sends 800 SRAMs against 800 targets. These targets are defended only by RV interceptors, and every SRAM reaches its target.

- **U.S. option C vs. Soviet option X:** 720 targets are destroyed. The United States sends 480 RVs against 240 targets defended by RV interceptors and 480 SRAMs against 480 targets defended only by RV interceptors; 240 + 480 = 720.

- **U.S. option A vs. Soviet option Y:** 1,200 targets are destroyed. The United States sends 1,200 RVs against targets defended only by SRAM interceptors.

- **U.S. option B vs. Soviet option Y:** 400 targets are destroyed. The United States needs 800 SRAMs to destroy 400 targets defended by SRAM interceptors.

- **U.S. option C vs. Soviet option Y:** 720 targets are destroyed. The United States sends 480 RVs against 480 targets and 480 SRAMs against 240 targets defended by SRAM interceptors: 480 + 240 = 720.

The key to all the calculations shown in the table was who had the last move at both the strategic and the tactical levels, and this was the point that Alain had overlooked. If the United States deployed 1,200 RVs and the Soviets had the last move at the strategic level, the Soviets would deploy only RV interceptors, and the outcome would be 600 targets destroyed. If the United States deployed 800 SRAMs (and, again, the Soviets had the last move at the strategic level), the Soviets would deploy only SRAM interceptors, and the outcome would be 400. Giving the Soviets the last move at the strategic level means that the Soviets would know what option the United States was pursuing and could react accordingly.

At the same time, I assumed that the United States had the last move at the tactical level and would know which targets were defended against what—RVs or SRAMs. But if the United States deployed 480 RVs plus 480 SRAMs, the outcome would be 720, regardless of whether the Soviets defended against RVs or SRAMs. Thus, by deploying a mix of RVs and SRAMs, the United States could ensure a level of destruction at least as high as 720 targets—an outcome that is better than that associated with either of the "pure" deployments. In short, the mix is the thing.

Dr. Enthoven was not persuaded by this analysis. But Dr. Brown ruled that the table fashioned by Dr. Enthoven was flawed because he failed to consider Soviet reactions. He thought that we should assume that the Soviets would have the last move at the strategic level, and so a mix of forces was the most prudent strategy to pursue.

So, for the time being at least, the triad would include bombers on "quick alert."

Later, as the Cold War began to wane and as other operational capabilities assumed a higher priority, the Soviets, in the presence of the ABM treaty of 1972, did not deploy an effective defense against ballistic missiles. All these reasons in concert operated toward the decision to stand down the bombers on alert.

While, in general, I preach that simplicity in analysis is preferred over complexity, in this case, my more-complex approach won. The lesson here is that one must not pursue simple approaches to the point that violence is done to the phenomena under examination. In particular, it is important not to treat the adversary as static. In military affairs, as in most fields of human endeavor, opponents react to each other's moves. Although this seems obvious, it is surprisingly common for advocates of certain policies or programs to assume that the adversary does not react to our initiative. In the case of Dr. Enthoven's comparison of Polaris and SRAM, this assumption was a fatal flaw.

Gaining Insight as to the Vulnerability of Submarines on Patrol

In the early 1970s, I served on a group directed by the Secretary of Defense to develop recommendations about the basing of our strategic nuclear forces: ICBMs in hardened silos, SLBMs in stealthy submarines, and bombers on alert at SAC bases. Among other matters, we were to examine and render findings as to the vulnerability of each mode of basing.

The Navy promoted the idea that missiles based on submarines at sea were quite invulnerable. To support this, they presented an analysis that had to do with Soviet attack submarines on patrol attempting to detect, track, and kill a Polaris submarine on station in the Norwegian Sea. According to their analysis—which used an expected-values approach to calculate the probability that a Soviet attack submarine would encounter, detect, establish track, maintain track, and kill the SSBN—the Soviet attack submarines were "never" successful.

Following the meeting in which this analysis was presented, I discussed the matter with Jasper Welch, by now a colonel. He suggested that we use their same inputs for values in the kill chain but determine the outcomes stochastically. That is, rather than multiply all the expected probabilities together, we should use a Monte Carlo simulation to determine the outcome (yes or no) of each event in the kill chain:

1. Soviet submarine detects Polaris: yes or no
2. Polaris submarine detects Soviet submarine and takes evasive action: yes or no

3. If "no" above, Soviet submarine tracks Polaris submarine: yes or no
4. Polaris submarine finally detects Soviet submarine and takes evasive action: yes or no
5. If "no" above, Soviet submarine engages and kills Polaris submarine: yes or no.

We ran many engagements. Once in a while, there was an unbroken series of "yes" answers for the Soviet submarine, and thus a "kill"— a successful engagement. We used the Navy's own probabilities in our Monte Carlo simulation to determine "yes" or "no" for each event in the sequence. The point here is that, if you try to achieve an event many times, you will finally succeed once in a while. Treating the outcomes as expected values tends to obscure this point.

In less than two weeks, amazingly, Colonel Welch had developed a computer model that ran these engagements in a stochastic manner. Once in a while, according to this analysis, the Soviet submarine was successful in detecting and engaging the Polaris submarine. Colonel Welch presented his work to the group. I introduced the item not so much to make the point that Polaris subs are vulnerable but rather to make the point that we should be cautious about moving everything to sea, as the Navy seemed to want.

Not surprisingly, the Navy representative on the working group was not pleased with our effort, our analysis, or our results. In fact, the next day, I was summoned by the Secretary of the Air Force, who had been called by the Secretary of the Navy, to explain why I was engaging in analysis of Navy systems. I felt that the Navy's objections to our work were way out of order. It was well known that the Navy was not shy about analyzing the survivability of Air Force systems, and that, in fact, they had maintained a group of analysts who episodically issued analyses underlining the vulnerability of ICBMs in hardened silos and bombers on alert.

The Navy's objections notwithstanding, we had made our point: The members of the group from OSD were impressed by the analysis and had a better understanding about the importance of avoiding a policy of putting all the nation's eggs in one basket.

Part of the Navy saw some merit to our approach. Colonel Welch related to me a month or so later that the Navy had decided that it should do its analysis stochastically and had quietly asked Colonel Welch to explain his computer program.

Even so, the Navy did not dismantle its analytical "hit team" and continued to render analyses of the vulnerabilities of missiles in hardened silos and bombers on alert.[9] But, for the moment, our analyses had stemmed the tide.

Defining and Promoting the Defense Support Program

An essential part of the framework for modernization that I used for many years (both when I was head of Development Plans at AFSC and when I was the head of AFSA) was the conceivers' action group (CAG). This, according to my concept, is a group of scientists, engineers, operators, and analysts charged with defining new concepts to achieve a particular operational objective. The idea was to have a core cadre of people and to augment this group with experts, depending on the particular challenge before them. In 1966 (or thereabouts), such a group was convened at the West Coast Study Facility under the direction of Lt Col Jasper Welch.

The challenge presented to the group was to define new and better concepts for providing early warning to our bombers that were on quick alert on many SAC bases.

The background and challenge were stated succinctly: On detection that the Soviets had launched ICBMs toward the United States, our bombers were to be launched to "safe havens" before the RVs from these ICBMs reached the bases at which our bombers were stationed. In the 1950s, the United States erected the Ballistic Missile Early Warning System (BMEWS) to provide this warning. The heart of this system was a set of three large radars, in Alaska, northern Greenland,

---

[9]   My annoyance at the Navy's practice of issuing such analyses was another reason I went after their SLBM survivability numbers. I had hoped to make them desist from criticizing our systems.

and northern Scotland. These radars would detect the ICBMs (or RVs deployed by these missiles) some 15 minutes after launch and approximately 15 minutes before they would reach the SAC bases.

There were several issues that created doubt as to whether we could reliably count on this system to accomplish the most critical function of providing timely warning:

- At best, it provided only a slim margin at the more northerly SAC bases. The bombers on alert would need almost the full 15 minutes to start engines, taxi, launch, and fly clear of the base.
- There was increasing concern that the Soviets might come up with a tactic, such as electronic jamming, to avoid detection.
- The system was vulnerable to false alarms, based on fluctuating conditions in the upper atmosphere, and Soviet SLBMs had shorter flight times.

So the challenge was clear: Define a better concept. More specifically, we needed to find a means of increasing both the reliability of the early warning system and to increase the time available for launching our bombers.

There was, at the time, a demonstration project called the Missile Defense Alarm System (MIDAS). This project was to demonstrate that it was feasible to detect the launch of an ICBM from satellites in LEO. The CAG believed that it was feasible to detect the plume of ICBM launches from satellites in LEO. The catch was that satellites in LEO do not dwell over any one region of the globe. To be sure that at least one satellite was on station over all regions in the Soviet Union where ICBMs were (or might be) deployed, the U.S. detection system would have to launch and maintain a constellation of dozens of satellites in orbit at all times. Without a constellation of this size, the Soviets would be able to exploit gaps in coverage. Such a large constellation was deemed to be prohibitively expensive.

On the other hand, one could cover all of the Soviet Union with just one or two satellites at the much higher geosynchronous orbit (GEO). But now there was the problem of detecting the plume from ICBMs from 22,000 nmi above the earth. The conventional wisdom at the time was that the technology of IR sensors would not provide detection at such a range. Thus, Colonel Welch felt that neither concept (LEO or GEO) seemed sufficiently promising to permit a recommendation to go forward.

Two engineers from Hughes Aircraft Company were among those involved in the sessions of the CAG. Back at their plant in Los Angeles, they focused their thoughts on how to develop better IR detectors. Their idea was to build detectors that would be optimized, or "tuned," to detect the particular IR wavelengths that were dominant in the plume of a liquid-propelled ICBM. They made some crude prototype detectors and conducted tests on the ground to verify their performance. Extrapolating from their test results (which used a very faint IR signal across a relatively short distance) to the problem at hand, they declared that it was indeed feasible to detect the plume of an ICBM from a satellite in GEO, using these specially tuned IR sensors.

Other engineers and scientists viewed the Hughes results with considerable skepticism. After all, the extrapolation was fairly heroic, to say the least. The Hughes engineers had put just two miles between their IR source and their sensor, while the operational system would have to detect plumes from a distance of 22,000 miles. But Colonel Welch vigorously followed this new lead. In about three months, he had the other scientists and engineers on board. These included engineers from the Aerospace Corporation and scientists from RAND.

Following a meeting with Colonel Welch, we decided the time had come to move out. Accordingly, I gained an appointment to see Gen Bruce Holloway, the commander of SAC. I had learned from past experience that the best path to getting something moving is to start at the top, and General Holloway sat atop the organization that would be the primary user of the data to be provided by this system.

The meeting with General Holloway convened in his office at 1700 or so on the appointed day. Just the three of us—Welch, Holloway, and myself—were present. Before the briefing, I cautioned Colonel Welch that he should avoid a discussion of technical details. For example, I did not want to get General Holloway involved in a lengthy discussion about whether the extrapolation of the results from the experiments that had been conducted was credible. The people best qualified

Box 4.2
Lessons from the Defense Support Program (DSP)

The best way to solve a complex operational problem is to convene an interdisciplinary group to tackle it.

Stating the challenge clearly is half the battle. Scientists and engineers, especially, are prone to get bogged down in explorations of their phenomenologies. They need clear direction to focus on a specific operational problem.

Related to this, concept development must be an exercise in problem solving, not an effort in advancing the state of the art of science or technology.

You may save a lot of money by taking a risk. The Air Force saved money by not undertaking other ways to solve the problem. By taking the risk inherent in starting a full-scale development program to implement a concept not fully demonstrated in terms of technical feasibility, we avoided expenditures on short-term fixes.

The decision to proceed came about considerably earlier than would otherwise have been the case because of the efforts of the CAG. Nowadays we focus on documents such as the MNS, the ORD, and the AoA, and proceed at the pace of a snail.

When you have devised a new concept, the place to "sell" it is to the most high-ranking person among the end-users, and the way to sell it is by showing clearly how it addresses an important operational objective for which he or she is responsible.

to make that determination had decided that the technical risks were acceptable and recommended, in light of the tremendous payoff from a successful effort, that a full-scale development program be started now. I saw little profit in spending the general's time revisiting issues of physics.

Despite my admonition to avoid technical details, the discussion soon devolved to just that. Suddenly, the general exclaimed, "I can make no judgment as to technical feasibility. You two would not be here making this presentation unless you believed that, in the presence of the expected payoff, the risk is acceptable." General Holloway went on: "This concept provides 15 minutes more warning than I now have with BMEWS. This means a lot to me. Today, I have to keep my aircrews out on stubs at the end of the runway. With an additional 15 minutes, I can put my crews in ready-rooms. I want you two to stay overnight and work with my staff in the morning on how to get this program started right away."

And so the program that became known as DSP commenced. The technology of the IR detectors progressed apace. Today, satellites in GEO regularly detect the plumes of even short-range tactical missiles.

CHAPTER FIVE

Modernizing Conventional Forces

Nuclear modernization programs were not the only ones plagued by a misuse of the term requirements. *In this chapter, General Kent relates how an irrational insistence that the contractor meet specified "requirements" almost ruined the program to develop and acquire the C-5 cargo aircraft. He also recounts his involvement in a number of other modernization programs—the F-15, the lightweight fighter, the AWACS, the Joint Surveillance Target Attack Radar System (JSTARS), and others—that constitute a major portion of the capabilities fielded by today's U.S. Air Force. In these accounts, he shows how well-crafted analyses, clear advocacy, and bureaucratic savvy can help steer complex programs toward successful completion.*

The C-5A Fiasco

The program to acquire the C-5A was (and remains) notorious for cost overruns and for underperformance. During the course of the C-5A's development, the Air Force was compelled to inject into the program large quantities of money beyond the amount originally budgeted. And more than once, important structural defects in the aircraft had to be repaired. The C-5A program is instructive chiefly as a means of illustrating how *not* to run a development and acquisition program.

The program was troubled from the start. Both contractors—Lockheed and Boeing—had submitted bids far below what was reasonable in terms of the cost of the program. They did this for business reasons: Boeing believed that, if it could win this contract, it could put Lockheed out of the business of making large transport aircraft—civilian or military. Lockheed, for its part, was determined not to be driven out of that business. Both bids were substantially below the estimated "should-cost" devised by the costing people at Wright Field. Nevertheless, in the late 1960s, the SPO signed a contract with Lockheed for the price it had bid.

There was a document that stated the "operational requirements" for the program, one of which was that the aircraft must have the capability for "rough-field takeoff and landing." It was declared early on by the engineers at Wright Field that the landing gear designed by Lockheed would not be sufficiently rugged to allow the aircraft to meet this "requirement."

Accordingly, the Air Force faced the decision to either amend this requirement or compel Lockheed to design new and stronger landing gear. General Ferguson, the head of AFSC, after considerable deliberation, decided to waive the requirement. After all, the C-5A was for strategic airlift and was not likely to be used for hauling cargo on a tactical basis. The smaller C-130 had that covered.

However, when General Ferguson's action became known, the Under Secretary of the Air Force sent him a stinging letter to the effect that General Ferguson lacked the authority to waive operational requirements. Rather, as the developer, his duty was to acquire systems that met these requirements. I had heard this frightening construction before in the MB-1 project. Reluctantly, General Ferguson withdrew the waiver and Lockheed redesigned the landing gear. This cost time and money.

Predictably, the new landing gear also increased the weight of the plane. Now a new problem arose. With the rough-field landing gear, the program could not meet another "requirement": The aeronautical manufacturer's planning report (AMPR) weight of the aircraft was not supposed to exceed a certain maximum.[1] Again, the question arose, "Should we waive this requirement?"

---

[1] AMPR (pronounced roughly like *ampere*) weight was as a measure frequently used in the 1960s and 1970s for estimating the cost of aircraft. It refers to the weight of the portions of

We conducted an analysis to gain some insight as to the effect of the extra weight. Comparing the performance of two aircraft—a C-5 that met the AMPR weight requirement and one that was heavier and did not—we found that the number of sorties needed to lift an Army Mechanized Division was exactly the same for both aircraft. The reason was simple: Both aircraft will generally fill up with cargo by volume before they reach their maximum gross takeoff weights. Said another way: Given that the C-5A loaded the outsized equipment of the division, the constraint was volume, not weight. So, operationally, there was no advantage in removing the extra weight.

The penalty for removing weight, however, was substantial and would be reflected in greater cost (to cover the costs of the design team) and slippage in the schedule. Moreover, in this case, to reduce weight, Lockheed would be forced to substitute titanium for steel, which further increased the cost.[2] So it seemed obvious that we should waive the AMPR weight requirement.

General Ferguson issued another directive to that effect. In no time at all, he received another letter from the Under Secretary of the Air Force to the effect that General Ferguson was not to concern himself with waiving requirements, and that the "waiver on weight" was null and void.

Another factor besides bureaucratic willfulness influenced the decision not to grant the waiver. The C-5 program was being conducted under a concept of Total Package Procurement (TPP), meaning that the contractor was responsible for producing a product that met all the contractual technical "requirements" for development, production, and support.

---

an aircraft that the aircraft manufacturer makes, as opposed to those purchased from other manufacturers and installed on the airframe. The AMPR weight of an aircraft, then, is the aircraft's empty weight minus such things as the wheels, brakes, tires, engines, instruments, batteries, and other items.

[2] Note the irony here. AMPR weight was inserted in the requirements in the first place because, as one aerodynamic engineer has said, "We buy airplanes by the pound." So, by controlling the AMPR weight, the Air Force had intended to limit the cost of the aircraft. But holding the contractor to a weight limit at this point in the development cycle, when the basic size and volume of the airframe were fixed, actually meant increasing the cost.

Lockheed had signed a contract to this effect. Now those with a stake in showing the validity of the TPP approach insisted that the Air Force show no flexibility with respect to its "requirements." "Hold their feet to the fire," they said. "Grant no waivers!"

General Ferguson protested that the goal should be to provide the Air Force with a large transport, not to bankrupt Lockheed. But his arguments fell on deaf ears. So, because some were more focused on flawed principles of acquisition than on a pragmatic evaluation of the realities of building an aircraft, Lockheed was obliged to take weight out of the aircraft, much of which came out of its wings. Predictably, the schedule slipped even more, and the cost went up. Worse still, the Air Force got a somewhat compromised aircraft out of the deal: Lockheed had to remove so much weight from the wing that the Air Force was obliged to rewing the C-5A early in its tenure in the inventory.

The primary lesson here is that blindly adhering to requirements can cause all sorts of serious problems. Had these detailed technical requirements been treated as performance parameters that we expected to achieve, the story of the C-5A might have been quite different. Another lesson is that management schemes, such as TPP, can have a pernicious effect on the development of systems when the people in charge become identified with such schemes to the point that showing the validity of their theories becomes more important to them than delivering a quality product to the service. Not long after the C-5 debacle, TPP was, mercifully, abandoned in favor of the time-tested "fly before you buy" approach. As "The Rationale for the Lightweight Fighter" (pp. 172–179) relates, this provided an opening for the Air Force to develop what became the F-16 fighter.

Defining the F-X and Saving the F-15

In the late 1960s, I was assigned to the AFSC. One of my problems at that time was to define performance characteristics for the F-X, the new fighter aircraft that was to replace the F-4. Personnel in Tactical Air Command (TAC) kept issuing statements that demanded greater performance. They wanted more payload, more range, a higher sustained turn rate, a higher instantaneous turn rate, greater specific excess power, a higher ratio of thrust to weight—and they wanted all these characteristics within a specified gross weight of the aircraft.

All these characteristics appeared in documents titled "Requirements," implying that the F-X must have them. The people at TAC were demanding an aircraft that simply could not be built. There are limits to every technology, and these "requirements" exceeded those limits.

I pointed this out to people at TAC, but to no avail. One can achieve any one of these specifications individually but not all of them in a single aircraft. If you want payload and range, you develop a high-wing-loading aircraft. If you want agility—the ability to turn and maneuver—you develop a low-wing-loading aircraft. You cannot have it both ways.

I recommended to General Ferguson that we wait until the "requirements" had run their course and then announce, with the concurrence of the commander of TAC, that there would be an "agonizing reappraisal." A group of colonels from TAC and AFSC would conduct this reappraisal and issue a final set of "requirements" (actually, performance specifications) that would lie within the limits of present technology. General Ferguson accepted this idea.

Not long after this, I was assigned to head AFSA. At AFSA, Lt Col Larry D. Welch was developing a computer-driven simulation called TAC Avenger that modeled one-on-one engagements with fighters.[3] One purpose of TAC Avenger was to see how various performance characteristics contributed to winning engagements. TAC Avenger was a very sophisticated simulation. It included fighter performance in five degrees of freedom: change in altitude ("y"), sideways motion ("z"), forward motion ("x"), pitch, and roll. The simulation maneuvered each aircraft as a skilled pilot would in the given situation. At that time, TAC Avenger was by far the best simulation of its kind. Colonel Welch

---

[3]   Colonel Welch had flown combat missions in F-4C aircraft over North and South Vietnam and Laos during the Vietnam War. After leaving my office, he graduated from the National War College and served in important posts in TAC. He was Vice Chief of Staff and commanded SAC before becoming the 12th Chief of Staff of the U.S. Air Force in 1986.

and two other officers were writing the computer code themselves, an amazing feat considering that they were all fighter pilots and had little training in computer science other than what they had received on the job.

I told these officers about the plan to reappraise the requirements for the F-X and told them to focus on this problem so that they could play a dominant role in the reappraisal. Specifically, they needed to be able to show the contribution of a stated performance characteristic to winning an engagement in the air. The engineers at the Air Force Research Laboratory told us what was technologically possible. The fighter pilots were to select (make the trade-offs) and define the "best" set of characteristics within the technology envelope set forth by the engineers.

In due time, General Ferguson called and told me that the time for the "reappraisal" was at hand. The group was about to convene, and he was authorized to name two of the members. He offered to allow one of these positions to be filled by someone from my Studies and Analysis office. I had told him previously that I had an officer who could be of great help.

"What's the name of this officer?" General Ferguson asked. I said that he was Lt Col Larry Welch. The general stopped me at once. "Look," he said, "this group is populated by colonels. I don't want to waste one of my slots on a lieutenant colonel."

"He will not be intimidated by colonels," I said. "I guarantee he will be better informed than anyone. He will dominate the group."

"All right, send this water-walker over here for me to talk to."

Right after his interview with Colonel Welch, General Ferguson was on the phone: "He's our man."

Just as I had predicted, Colonel Welch dominated the group. No one else had anything remotely like TAC Avenger. The reappraisal defined characteristics for the F-15 fighter, which also became an effective attack aircraft as the F-15E.

The program prospered, and the F-15C and F-15E became the backbone of the fighter force in the Air Force. The F-15C was the premier fighter in the world for a long time.

This story suggests some sound principles:

- Engineers should define the limits of technology.
- Experienced operators should define the best balance of characteristics within those limits.
- Those characteristics then define the system.

Some time later, leaders in OSD raised an issue: Why should there be separate programs to produce what appeared to be similar aircraft, the F-14 for the Navy and the F-15 for the Air Force? A study group was formed, headed by Dr. Alan Simon of the office of the Director of DDR&E. Some people within OSD wanted to terminate the F-15 program to save money. To keep the program, we would have to demonstrate how the F-15 was far superior in air-to-air combat to the F-14.

Colonel Welch did runs with TAC Avenger pitting both the F-15 and the F-14 against first-line Soviet fighters. The simulation showed that the F-15 was far superior as a fighter. Because of its longer range, the F-14 was better at intercepting Soviet bombers when they attacked the fleet, but it was not nearly as capable as the F-15 in combat against Soviet fighters. The F-14 had more payload range, but the F-15 was far superior in agility.

The Navy launched a public relations effort that made newspaper headlines: "The F-14, the World's Greatest Fighter." I told the admiral in the group that, according to our calculations, this statement was far from true, indeed, that the F-15 was much superior to the F-14 as a fighter. The F-14 made the trade between payload-range and agility in favor of payload-range. The Air Force made the trade in favor of agility. The admiral responded that he based his claim on work that had been done at the Center for Naval Analysis, where he was vice commander. The center briefed their work to the study group, and I saw to it that Colonel Welch was there.

Back in my office, Colonel Welch expressed confidence that we could easily substantiate our claim that, as a fighter, the F-15 was far superior. "Their work pales in comparison to ours," he said. He felt certain that he could make a convincing presentation to the study group, and he did just that. Even the admiral was impressed. At this point, I said that, if there was only enough money for one program, it should be the F-15 with a naval variant.

Dr. Simon was very informed about aeronautical engineering and was a very astute individual. He was convinced that the Air Force was right, that the F-15 was by far the better fighter. In due time, he announced that there would be two programs.

This episode demonstrates the power that analysis based on a brilliantly conceived simulator (in this case, TAC Avenger) can have. It also shows the importance of anticipating the emergence of issues in the debate: Knowing that the F-X program would inevitably face the need for an agonizing reappraisal of "requirements" and that this was an issue of core importance to the future of the Air Force, we invested considerable time and talent in understanding everything we could about air-to-air combat, the technology available for the next-generation fighter, and the trade-offs available in aircraft design. These same investments helped us show the poverty of the Navy's arguments in favor of its interceptor.

There was one disturbing aspect of TAC Avenger: In simulations of combat against Soviet-designed fighters, it showed remarkable kill ratios, generally on the order of 15 to one or more, which some might think unrealistic. I cautioned my people not to advertise a kill ratio of more than 15 to one, lest people doubt the validity of the simulator. But perhaps I was too hasty. The F-15 has never lost a fight in actual combat, due both to the superior characteristics of the aircraft and the high quality of training given to its pilots.

The F-15 became an extremely successful aircraft. F-15Cs assured air superiority in the 1999 Kosovo conflict, in the 1991 Iraq war, in the 2003 Iraq war, and whenever else they fought. The F-15E, adapted for ground attack, has also played an important role in these conflicts, vastly improving the Air Force's ability to destroy an enemy's fielded forces.

The Rationale for the Lightweight Fighter

In the early 1970s, a group of analysts led by then–Maj John Boyd defined an aircraft known as the lightweight fighter (LWF). This aircraft was to cost about 60 percent as much as the cost of an F-15 and was promoted as being 70 percent as effective as the F-15. Thus, in terms of a standard cost-effectiveness analysis, there was a slight advantage in favor of the LWF.

Major Boyd came to my office and briefed me on the concept of the LWF. I had no qualms with his claim that the LWF could be 70 percent as effective as the F-15—with one important caveat: Boyd's statement applied only to air-to-air combat. It did not apply when the aircraft was employed in the ground attack mode. For the attack versions of both aircraft, the ratio of effectiveness was apt to be more like two to one in favor of the F-15, in large measure because the F-15, being a larger, twin-engine aircraft, would have about double the bomb load of the LWF. But defensive counterair was an important mission for the Air Force at that time, since we faced large numbers of Soviet and Warsaw Pact aircraft in Central Europe, so the focus on air-to-air combat by the proponents of the LWF was relevant.

In their advocacy of the LWF, Major Boyd and his followers seemed driven not only to extol the virtues of this new aircraft but also to advance the theory and practice of air-to-air combat. Major Boyd, for one, had the credentials to do so: He had been a fighter pilot and had a record of success in air-to-air engagements. Unfortunately, Boyd and company conducted their advocacy in a way that denigrated supporters of the F-15, which at that point was well along in the development process. Not surprisingly, this alienated many of the leaders of the Air Force, for whom developing and fielding the F-15 was a top priority. But Boyd and his acolytes were on a mission, and they could not be persuaded to alter their pitch.

The Vice Chief of Staff, Gen J. C. Meyer, called me into his office after a staff meeting. He had in hand a memo from Lt Gen Otto Glasser, then Deputy Chief of Staff for Research and Development. The memo stated that General Meyer should hear the briefing about the LWF. Before he sent his memo to General Meyer, General Glasser and I had talked about this matter and were of the same mind: We wanted General Meyer to hear about the characteristics of the new fighter the group had defined. General Glasser had even asked the engineers at Wright Field if the performance features claimed by Boyd were possible in the smaller aircraft. They attested to the technical feasibility of the aircraft.

Upon receipt of the memo by General Glasser, General Meyer told me that he would agree to hear the briefing if and only if I recommended that he should. He also made it clear that he would prefer not to have to listen to the briefing, having been irritated before by the Boyd group and its strident campaign for the LWF. In fact, General Meyer stipulated that he required a written memo from me stating my view that he should hear the briefing before he would agree to do so.

Soon, Maj Everest Riccioni, one of Major Boyd's followers, was in my office with the LWF briefing in hand. After a lengthy discussion in which he presented the briefing to me, I stated that I would urge General Meyer to hear *a* briefing about the LWF, but not the briefing Riccioni had just shown me. "There are two distinct parts to your briefing," I said. "The first part states that technology marches on and the Air Force can have a fighter with impressive performance at 60 percent of the cost of an F-15. The second part of the briefing alleges that those who support the F-15 lack a basic understanding of air-to-air combat. I will recommend that General Meyer receive a briefing that sticks religiously to the first part and contains not a hint of the second part. General Meyer supports the F-15, and he needs no instruction from you (or anyone else) about the practice of air-to-air combat. After all, he was the leading American ace in the European campaign in World War II."

Major Riccioni protested. I pointed out that he was negotiating from jail. The easiest and least risky course for me was to tell General Meyer he should not hear the briefing. I insisted that I would only endorse a briefing that reflected my view of what was constructive, repeating that it would convey only material from part one and would not include a hint about part two. In time, Major Riccioni saw that he was in no position to argue, and together, he and my staff developed such a briefing.

I wrote a note to General Meyer and urged him to hear the briefing. He promptly made it known to me that I had failed him. "All right," he said. "I will hear what this major has to say. But I hold you responsible for the whole affair." A date was set.

Late in the afternoon the day before the appointed date, I was called out of town. Every instinct told me to cancel the briefing, but it was hard to get on the calendar of the vice chief, so I did not call to cancel. I did call Lt Col Larry Welch (who worked for me) and Major Riccioni to my office. "I trust you, Major," I said. "I won't be there but I trust that you'll stick to the script we have developed: just part one, nothing from part two. Do not even take those other charts in your briefcase." Major Riccioni agreed.

I told Colonel Welch that I would call him at his home when I returned the next day. When I called him I asked, "How did the briefing go?"

"It was a disaster," Larry replied. "The major had hardly gotten into the briefing when a statement by General Meyer prompted Riccioni to exclaim, 'It's clear that General Kent was wrong and that a review of the fundamentals of air-to-air combat is necessary.' He took some charts out of his briefcase and was barely into the material on the first chart when we were dismissed."

My worst fears had been realized. General Meyer was, to say the least, not pleased. He asked me if I had bothered to hear the briefing. I replied that I had but added that I could not control Major Riccioni's behavior when I wasn't there. General Meyer announced that he was sending Major Riccioni to Korea so that he would be in no position to muddy the waters. He probably felt it was not necessary to add that this was the last he wished to hear about the LWF.

Major Riccioni, to his credit, came to my office to apologize. He acknowledged that he did not adhere to our agreement and regretted that he had blown an opportunity to advance the cause of the aircraft in which he believed. In later years, we became good friends, and his career as an aeronautical engineer prospered. In time he was a full colonel and chief of the Flight Dynamics Lab at Wright Field.

Notwithstanding the Riccioni-Meyer debacle, Boyd and company continued to promote the idea of the LWF, but they gained little traction. Then an unexpected opportunity appeared. General Glasser called me to his office. "David Packard, the Deputy Secretary of Defense," he said, "is interested in promoting the use of prototyping as a means of invigorating the defense acquisition system. He wants to jump-start this effort with some demonstration projects, and he has requested that each service recommend one new system that they would like to prototype. He will provide the funds."

General Glasser continued, "There is a meeting about this tomorrow called by Dr. Grant Hansen, the Assistant Secretary of the Air Force for R&D. At that meeting, I will suggest that the Air Force respond with the LWF as our candidate. Dr. Hansen knows that General Meyer is not a fan of this concept, and he may be reluctant to accept my suggestion. Grant Hansen knows that the vice chief likes you, so it is your job to convince Hansen that General Meyer can be persuaded to support the demonstration. We'll emphasize that this is a 'prototype' effort, not a full-scale development program; there is no commitment to develop and acquire the system."

I attended the meeting and made the pitch that, while General Meyer had expressed opposition to the LWF and its proponents, his real problem was not with the concept of an LWF per se, but, rather, with the way the concept was being promoted. "If the LWF were cast as a complement to the F-15 and not as a substitute for it," I said, "he probably could be persuaded to support it." "It would help," I added, "if those who favor the LWF would stop acting as if they were the only people in the nation who understood anything about the nature of air-to-air combat."

Hansen wasn't buying it. He did not believe that General Meyer could be brought around on this issue, and he was unwilling to proceed to recommend to OSD that we prototype the LWF until the general actually was converted.

General Glasser then urged me to go to General Meyer myself and make the case for the LWF. I asked Colonel Welch to prepare a ten-slide briefing making the case that the Air Force would not be able to afford enough F-15s to replace the F-4 on a one-for-one basis and that a "high-low mix" concept would be essential. I spent the next few days pondering whether and how to take on this daunting assignment. Before I resolved the matter, General Meyer called me to his office. "I see that people are talking about offering the LWF as a candidate for prototyping," he said. (Apparently, the rumor mill had gotten hold of

General Glasser's idea.) "Do you have a better rationale for the aircraft than I have heard to date?"

Here was my chance. I told General Meyer that I had thought more about the notion of a high-low mix of air-to-air fighters and how to quantify this concept. Fighter aircraft, I said, like most major combat systems on the battlefield, act as both shooters and victims. That is, they have the potential to kill the enemy's fighters, and they are also potential victims of enemy fighters. This being the case, Lanchester's square law applies. That is, the number of systems one commits to an engagement scales as the square, while their effectiveness per aircraft scales as the first power (see Table 6.1). "If we were to acquire LWFs rather than more F-15s, we could have, for the same amount of money, 1.67 times as many F-16s as F-15s." This is because the LWF was said to cost about 60 percent as much as an F-15. "1.67 squared is 2.8. And 2.8 times 0.7 comes to about 2." I used 0.7 because the LWF's proponents claimed that their aircraft would be 0.7 times as effective as the F-15 in air-to-air engagements on a unit-by-unit basis. "Therefore, investing in the LWF as part of a high-low mix would provide roughly twice the force effectiveness as an investment in F-15s"—this as far as air-to-air combat was concerned.[4]

General Meyer listened carefully, and in the end, he agreed that General Ryan, the Chief of Staff, should see Colonel Welch's briefing. When we showed the briefing to the chief, he complained, not because we were advocating a concept that competed with the F-15 but rather because he had not been presented with the concept of a high-low mix and the LWF earlier. The result was that the Air Force did recommend that the LWF be developed as a prototype. The prototype program became, as General Glasser intended all along, a full-scale development program that ultimately yielded derivatives of the F-16 for the Air Force and derivatives of the F-18 for the Navy. Both aircraft have been highly successful; derivatives of both are still being produced today.

A mythology of sorts has grown up around John Boyd and his followers and how they outmaneuvered the corporate Air Force into

---

[4] For another application of Lanchester's square law to a question of fighter force structure, see "Another Episode with SABER GRAND," pp. 217–223.

Box 5.1
Lanchester's Square Law

Let

B = the number of Blue agents (in this case, aircraft)

R = the number of Red aircraft

γ = the effectiveness of each Blue aircraft

λ = the effectiveness of each Red aircraft.

Lanchester's square law states that we have equality when

$$\left(\frac{B}{R}\right)^2 \times \left(\frac{\gamma}{\lambda}\right) = 1.$$

This is the mathematical embodiment of the principle that "quantity has a quality all its own." So, if Red aircraft outnumber Blue aircraft by a ratio of, say, 2:1, each Blue aircraft must be four times as capable as each Red aircraft for the fight to be even.

For example, if B equals 20 and R equals 40,

$$\left(\frac{B}{R}\right)^2 = \left(\frac{1}{2}\right)^2 = \frac{1}{4}.$$

Then, if

$$\left(\frac{1}{4}\right) \times \left(\frac{\gamma}{\lambda}\right) = 1,$$

it follows that

$$\frac{\gamma}{\lambda} = 4.$$

building the F-16.[5] The reality, as I view it, was rather different and more nuanced than that suggested by the myth. Few people today know of the crucial role of Colonel Welch and General Glasser in getting the program started.

Demonstrating the AWACS

In the late 1960s, the Air Force advanced the concept of equipping a large aircraft with a highly capable moving-target indicator radar so as to detect and track low-flying aircraft from a station between 20,000 and 30,000 feet in the air out to ranges approaching 100 miles. Conventional land-based radars could not detect low-flying aircraft beyond a few miles because of ground clutter at the horizon. Finding a way to elevate the radar was essential to defeating attacks by low-flying aircraft. The key technology for attaining the capability to "look down" was to get some 50 dB enhancement of the signal from the enemy aircraft by Doppler processing to detect and track enemy aircraft against a background of the earth's surface.[6]

Dr. Tony Batista was then a key staff member of the House Armed Services Committee. Tony was smart, and he worked hard. When he took a position on an issue, he was usually correct. He therefore carried considerable clout when it came to decisions about funding new concepts.

---

[5] Boyd's biographer, Robert Coram, refers to the LWF as "one of the most audacious plots ever hatched against a military service." He goes on to claim that the program was "done under the noses of men who, if they had the slightest idea what it was about, not only would have stopped it instantly, but would have orders cut reassigning Boyd to the other side of the globe" (Robert Coram, *Boyd: The Fighter Pilot Who Changed the Art of War*, New York: Little, Brown, and Co., 2002, p. 245).

[6] Doppler processing takes advantage of the phenomenon that energy returned from a moving object incurs a change in frequency proportionate to the closing velocity of that object relative to the radar, just as a train whistle will change pitch depending upon whether the train is coming or going relative to the listener.

A radar engineer had convinced Tony that the technology of signal processing needed to make the AWACS feasible was not at hand, so Tony stated that he would oppose funding for the new concept.

The Air Force had conducted some tests; however, the results were not convincing—especially when the burden of proof was on the Air Force and in the face of opposition by prominent scientists and engineers elsewhere. Tony was adamant that we had to have a more convincing story. Dr. Al Flax, the Assistant Secretary of the Air Force for Research and Development, issued a directive that a demonstration project would be conducted under the direction of General Kent. I was at that time head of Development Plans at AFSC at Andrews AFB.

With the help of Dr. Harry Davis (who worked for Dr. Flax), I defined such a project. Key elements of the demonstration project included the following:

- The Air Force would equip a large aircraft with an antenna on the side that would enable it to "look" down.
- The aircraft would also be equipped with a polar projection indicator (PPI) scope. This would display radar returns after they came through the signal processor.
- Relevant contractors with expertise in radar and signal processing technology (maybe about six) would be invited to send teams of technical experts to Wright Field to hook their radars and signal processors to the antenna fixed on the large aircraft and to the PPI scope in the rear of the aircraft. Each contractor would be paid a flat fixed fee intended to cover the expenses of participating in the effort.
- Personnel at Wright Field would arrange the schedule according to which the contractors would appear at Wright Field.
- The Air Force would make it clear from the outset that this was a demonstration project, not a development program. The results of the project would have no bearing on subsequent decisions about source selection. We rigorously maintained that the project had a simple and limited purpose: to provide overwhelming evidence of the feasibility of the technology of Doppler processing to

reject ground clutter (and, in the process, to change the mind of Dr. Tony Batista).

- Our demonstration was, in the first instance, to be focused on detecting large transports (commercial aircraft) against a background of cornfields in the area around Dayton, Ohio. (This reflected a walk-before-you-run approach. Detecting large aircraft against a flat surface is easier than detecting small aircraft over, say, the Rocky Mountains.)

- The contractors were told to "come as you are." The money to be paid to them was intended to cover only the expenses associated with hooking up an existing radar and processor to our antenna and PPI scope. The money was not for maturing the technology by the contractors. If a contractor did not have the technology at hand, so be it.

Dr. Flax approved this approach, and I went to Wright Field to explain and implement it. Much to my dismay, the approach I had defined was greeted by the people at Wright Field with considerable skepticism, if not downright hostility. The hostility arose in large measure from the "not invented here" syndrome. The engineers at Wright Field resented this outsider, who was not a radar engineer, defining in some detail how they were to run their business. In their eyes, what they had been doing was just fine. They saw no reason to change. My response was that they were proceeding at a snail's pace and that changing Dr. Batista's mind would not come about for years. The leaders of the Air Staff wanted faster progress, and I believed that Dr. Davis and I had devised a reasonable, albeit unconventional, approach to quickly demonstrating that the technology was at hand, at least with regard to detecting large aircraft.

I had stated that we should commence immediately and be ready for the first contractor to arrive at Wright Field in about six weeks. A colonel in the procurement office at Wright Field said that such an approach was tantamount to letting a sole-source contract, which he was not about to do. It would take at least a month or so to develop the request for proposal (RFP) and another month or two to go through the process of source selection. I patiently explained to the colonel that our plan called for letting all relevant contractors participate in the demonstration and that each one would get an equal amount of money. As such, the approach could hardly be characterized as a sole-source contract. I reiterated that we did not intend to waste time negotiating a price for the work of each contractor. We would offer a flat fixed fee, period. We intended to set the fee at a level that would cover the cost of hookup. If a particular contractor did not want to participate, so be it. My pristine logic fell on deaf ears.

I finally played my trump card. I pointed out that it was a close call whether to have Wright Field or the Rome Air Development Center in Rome, New York, conduct the project. If the people at Wright Field felt they could not, or should not, conduct such a project, I would be off to Rome that afternoon. At this juncture, Dr. Fred Orazio, a senior engineer at Wright Field, marched to the front of the room. He announced to me and to his colleagues in the room that he would support the Kent-Davis approach. "Yes, General Kent, this project is somewhat at odds with the way we have been doing business. But for the purpose intended, it is right on. We will conduct the project in the manner you have defined."

As I recall, six contractors participated. All six demonstrated that it was possible to detect low-flying aircraft (even aircraft smaller than large transports) against ground clutter over the terrain of Ohio. The demonstrations were completed within less than three months of my trip to Wright Field. Not long thereafter, Dr. Batista, bless his soul, in the presence of the results from this project, graciously agreed that the necessary technology probably was at hand and that the Air Force should proceed with a full-scale development program.

The rest is history. A bevy of combat units in the United States, NATO, and elsewhere are equipped with AWACS today (although the program encountered a "near-death experience" before it reached IOC; see "Strategies to Tasks: A Construct for Advocating New Concepts," pp. 115–121).

This case offers several lessons for those engaged in developing and acquiring systems:

Box 5.2
The Development of the Airborne Warning and Control System

When we organized the technology demonstration effort at Wright Field, I was clear in specifying that the objective was to show that we could detect and track large aircraft from an airborne radar. There were two reasons for this: First, obviously, it is easier to detect and track an object with a large radar cross section than one with a smaller one, and I was eager to have a success. Second, and more important, the focus on large aircraft was justified operationally by our belief at the time that the primary mission of the AWACS would be to detect Soviet bombers approaching the United States and to vector interceptors against them.

As the program approached IOC, however, the AWACS mission evolved toward that of a control system for theater warfare. For example, NATO planners looked to AWACS to provide a survivable means for controlling large-scale defensive air operations in Central Europe. This mission would require the radar on AWACS to be able to detect and track small, fighter-sized aircraft. The radar cross section of these aircraft is about 20 dB lower than that of bombers or transports. Westinghouse, the contractor selected to develop and produce the radar, had to undertake fairly heroic efforts to achieve this capability. It did this by designing an antenna with ultralow sidelobes—an achievement that advanced the state of the art considerably and, in fact, was not surpassed for decades thereafter.

- Facing the task of advancing the ball on AWACS, my focus was on people and the decisions they made, not on preparing documents or filling squares in a faceless bureaucratic process. We could have spent a year or so (maybe longer) preparing the documents called for in the 5000-series regulations—ORDs, MNSs, milestone decision documents, and so on. But even if these were all perfectly executed, they would not have put rubber on the ramp until one man—Dr. Tony Batista—was convinced that the requisite technology was in hand (or nearly so) and that there was no undue risk in starting a full-scale program to develop and acquire the system proposed. That was the problem, and it was the problem we worked.

- It is vitally important to distinguish between demonstrating a technology and developing and acquiring a system. Often, contractors who are successful in demonstrating some new technology will try to blur this distinction and argue for a "seamless transition" to program development. This is wrong from both the standpoint of being legal and from the standpoint of being logical. Legally, source selection for a major program must be the result of a decision by a high-level official, sometimes the Secretary of the Air Force. Generally, the contract must be competed. Failing to do this will result in challenges to the program and delays. And factors going beyond simple mastery of a technology, notably, the ability to produce, must be taken into account in awarding a major contract.
- People at government labs are often more interested in maturing technologies than in putting rubber on the ramp. They award contracts to demonstrate—not to develop and produce. This is all right up to a point, but when a decision is made to develop and field a new major system, their focus must shift from advancing the state of the art of a technology to designing something that can be produced and deployed.

Starting the JSTARS Program

In 1974, I retired from active duty with the Air Force. A few years later, Dr. Ken Perko, head of the Tactical Technology Office of the Defense Advanced Research Projects Agency (DARPA), called me, wanting to talk to me about an important matter. He told me of a technology project in DARPA that had demonstrated an amazing radar that cleverly exploited the Doppler effect. This radar could detect and track moving vehicles at ranges up to 100 miles, even when they were moving at only a few kilometers per hour.

Dr. Perko said that DARPA had not been able to interest the higher levels of the Air Force in a program to develop and acquire such a radar. Dr. Perko wanted approval from Gen Robert J. Dixon, the commander of TAC, but Dixon was extremely reluctant to join any endeavor involving DARPA. He had recently worked with DARPA on another concept and was quite unhappy with DARPA's performance. Dr. Perko was aware that I had known General Dixon for many years and that we got on quite well, so he wanted me to approach General Dixon on this matter. I told Dr. Perko that I would look into this proposition and get back to him.

The radar he described sounded very promising. If it could perform as he stated, it would significantly improve our ability to find and target fielded Warsaw Pact ground forces. Instead of having to wait to engage these forces in close combat, we could detect them at considerable distances (up to 100 miles away) and strike them before they could engage our forces on the ground. A radar like this could greatly increase the effectiveness of airpower in interdicting enemy ground forces.

A week later, I met with Dr. Perko and laid out the following approach. First, we would "read" Lt Gen Robert T. Marsh into the project. At that time, General Marsh was the commander of the Air Force's Electronics Systems Division at Hanscom Field in Massachusetts. I expected that General Marsh would ask the engineers at the Rome Air Development Center, which was under his direct command, to examine the technical feasibility of such a project. Once they had assured General Marsh that the concept was feasible and he was convinced, I would seek an audience with General Dixon to recommend a joint DARPA–Air Force program. The program would be designed to ensure a smooth transition from DARPA to the Air Force, which would ultimately develop and acquire the system. I envisioned a conference of the various contractors at Hanscom under the auspices of General Marsh. The meeting would be at Hanscom to underline the point that the Air Force was indeed in charge.

DARPA would provide 85 percent of funding for the first year of the project, 50 percent for the second year, and 15 percent for the third year. Thereafter, the Air Force would provide all the funding. Also, General Marsh would be the source-selection authority.

Dr. Perko was reluctant at first to agree to all these conditions, but I refused to see General Dixon until they had been met. Eventually, he agreed, and I contacted General Marsh. Within a week or so, General Marsh called back and said that his engineers had assured him that the radar was technically feasible and that he would advocate a program to develop and acquire it. Thereupon, I called General Dixon's office at Langley AFB and asked for an appointment.

"What is the subject?" asked his scheduler.

"I cannot say over the phone," I replied. He relented and gave me the appointment.

On the appointed day, I made my way to Langley. For a few minutes, General Dixon and I talked about the "good old days," when we were colonels in the Plans Section of the Air Staff. Finally, the general asked, "What brings you here?"

I started by saying that "controllers" orchestrating air attacks on enemy ground forces would profit greatly if they could detect and track moving vehicles, such as tanks and other vehicles, at great distances, perhaps up to 100 miles away. A system of this kind would do for the mission of attacking enemy vehicles moving on the ground what AWACS does for attacks on enemy aircraft flying through the sky.

General Dixon saw the utility but wondered whether such a system was possible. "Where is this magic radar? I never heard of it before! What have you been smoking?"

I replied that General Marsh was ready to testify that the concept was feasible.

"How does General Marsh know what you just told me?" asked General Dixon.

"Because of prearrangements, sir." I replied.

"Then there is no need to call him! I trust you as the messenger. You always do your homework!"

Now was the time to tell General Dixon that the whole concept had its beginnings in DARPA. I drew a deep breath and told him, expecting that he would explode.

"You are their spy!" he exclaimed.

"I prefer to think of myself as perhaps a double agent," I replied. "I have laid out an approach whereby the program will transition smoothly from a DARPA technology project into an Air Force development program. Both the Air Force and DARPA will surely benefit if we go ahead." I explained the approach and conditions that I had previously outlined to Dr. Perko.

Box 5.3
Implementing the Joint Surveillance Target Attack Radar System

The account of events in this chapter has been streamlined considerably to make the point about how to go about getting programs started. That account leaves the impression that there was a smooth ride for JSTARS from inception to IOC. Such was not the case. The original concept by Dr. Perko of DARPA involved both the finder (the radar) and the shooter (a missile of some sort). He gave the name "Assault Breaker" to the overall concept. After some thought, he decided to have the Air Force develop and acquire the finder and to have the Army develop the shooter. Thus, the focus in the Air Force was on deploying the radar.

Initially, the program to develop the radar was called "Pave Mover." The program proceeded rather smoothly through the selection of the contractor, Norden, but then someone in OSD decided that the program should be "joint." (In the late 1970s, *joint* had already become the watchword.) I argued that the key element was that the system could (and would) be employed in joint operations. For example, if the finder reported the location of some target(s) to a "joint engagement control center," the system would, by definition, be used jointly. Whether the system was developed and acquired by a joint program was not relevant. They key term is *joint operations*, not *joint development*.

This deathless logic fell on deaf ears; the program became "joint," and JSTARS was born. The contractors who lost in the source selection for Pave Mover were delighted to have a second chance, but to no avail. Norden also won the second time. The decision to make the development program "joint" cost time and money with no value added.

The shooter part of Assault Breaker languished for some time. Finally, the Assault Breaker concept was realized when the Army fielded its Army Tactical Missile System (ATACMS), though the sophisticated anti-armor submunitions originally intended for the missile encountered numerous development problems and were never fielded in large numbers. Nevertheless, JSTARS has proven to be a valuable means for commanders to gain awareness of the disposition of mechanized forces over a wide area. Information provided by JSTARS is fed to shooters of all types, including Air Force interdiction sorties.

"All right!" General Dixon declared. "I support the program you have described on the condition that you personally will see to it that DARPA adheres to the agreed approach—no exceptions." I agreed, and the deal was done. My leverage was clear. If DARPA did not adhere to the stated conditions, I would be obliged to report this to General Dixon, and he would probably withdraw his support.

The program began, and today we have the Joint Surveillance Target Attack Radar System (JSTARS). I helped start this program by serving as an "honest broker," by delineating certain rules and conditions and getting both sides to agree to them. Today, Air Force officers still have difficulty transitioning systems from DARPA projects (intended to demonstrate certain technologies) to Air Force programs (intended to develop and produce weapon systems). They might consider using the approach we took with respect to JSTARS.

Several other lessons emerge from this story:

- Effective advocacy begins with showing the need for a new capability.
- It pays to have friends in high places. As the person charged with organizing, equipping, and training units within the Air Force for theater warfare, General Dixon was in a position to get the ball rolling quickly toward an important new capability. Because of our longstanding personal relationship, I was able to bring this opportunity to his attention directly without having to go through layers of subordinates.
- Anticipate the decisionmakers' key questions and address them before they are even asked.
- Important decisions often revolve around personalities more than formal documentation. General Dixon did not demand any formal document declaring that such a system was technically feasible. General Marsh's assurance was enough for him to make the decision to proceed.

Keeping the Global Positioning System Alive

In the early 1970s, I was the head of the Weapon System Evaluation Group (WSEG). This was a joint group—Army, Navy, Air Force, Marines. I had two bosses, the chairman of the JCS—then ADM Thomas Moorer—and the Director of Research and Development in OSD, Dr. John Foster.

Dr. Foster called me to his office one day and related, "The day after tomorrow, there will be a meeting in my office about a project known as NAVSTAR [Navigation Satellite Timing and Ranging]. This system is designed to allow a person to determine his position on the globe by listening to signals from four different satellites and observing the time of arrival of a stated pulse at his receiver.[7] The time differences of arrival can be used to calculate the position of the receiver within several meters. My problem is that no one, and I mean no one, is willing to put money in their budget to continue the project." Dr. Foster continued: "I need at least one voice at the meeting saying that we should continue the project. I know you are familiar with the project because they worked on it at your West Coast Study Facility a few years ago." Dr. Foster then came to the point: "Can I count on you?" My answer was "yes."

I did some hasty research. I recalled the time some three or four years earlier when I first heard of this concept. A Dr. Noika of Honeywell was temporarily attached to the West Coast Study Facility in the Los Angeles area. He told me of the concept for locating your position on a reference sphere by observing the difference in time of arrival of the signal from three or more satellites in LEO. At the time, I pointed out that, given the speed of light as approximately 1,000 feet per microsecond, we would have to be able to measure time accurately down to less than 1 microsecond if the system was to be of use in locating combat units on the ground. He said he understood. I asked, "How are you going to do that?"

He replied, "I do not know for sure. But they are working on it. Some day they will succeed."

Some years later, time resolution of a few microseconds had been demonstrated. This was good enough for locating your position in the context of navigation over the ocean to bring you to a destination like Ascension Island. In fact, the principal use touted by the people on the project was in the C-5 transport aircraft. But there was no traction

---

7 Measurements from four satellites are needed to solve for receiver clock error and eliminate it from the position solution. It is conceivable that three satellites could be used to solve for receiver clock error and a two-coordinate position on a "reference sphere."

there. The C-5 already had a triple-redundant inertial system, and this was supplemented in most areas of the world by ground-based radio navigation aids, such as LORAN (which stands for Long-Range Aids to Navigation). So improving point-to-point navigation for long-range aircraft was not, in itself, a compelling reason to deploy NAVSTAR. The best I could do by way of generating a rationale for the system was that it could be used by combat units on the ground to determine their positions (often a more difficult challenge than it seems) and for maintaining the location of aircraft, such as AWACS, that need to "orbit" over specific areas for extended periods. My vision of the application of this new technology was seriously clouded by the idea that the time resolution was around a few microseconds at best. This meant that NAVSTAR's resolution would be no better than a few thousand feet.

At the meeting, none of the representatives of the various organizations present were interested enough in implementing the concept to put money in their budgets to do so. The representative from the Air Force was one of the most skeptical. He stated over and over that "the Air Force has no requirement for this system." I suggested my ideas but neither the Air Force nor the Army found them compelling.

The meeting finally came to a dismal end. I stayed behind to review the bidding with Dr. Foster. He thanked me for my effort—even though I had accomplished little. I did get one thing out of the meeting, however, and that was confirmation of my belief that characterizing the system as an aid to navigation of ships or aircraft was a loser. There was no real need to improve upon the systems already available. Accordingly, I suggested to Dr. Foster then that, if we were to continue the project, the key word should be *position*—the position of combat units on the ground, in the air, or at sea. Also, we might be able to use the system to guide long-range cruise missiles to a "basket" or general area. Dr. Foster quickly agreed and decided that the name of the concept henceforth was the Global Positioning System (GPS).

We all owe a debt to Dr. Foster, who found the money to continue the project despite the total lack of interest from the potential users. The project continued to have its ups and downs with regard to funding. But Brad Parkinson, who was the leader of the project, made steady progress. We owe him and engineers at Aerospace Corporation for defining and implementing the masterful architecture of the system.

Finally, the idea of a GPS caught on. Today, no combat unit, down to the individual soldier, would think of leaving home without it. Because the GPS signal is freely available to all, the technology is also being used for a host of civilian applications, from in-car navigation systems to cellular telephones.

This story is worth telling because, again, progress often entails risk. In this case, the risk was both technical and operational. Everyone greatly underestimated the potential utility of a future technology that would enable us to measure time down to a resolution of a few nanoseconds and, hence, to determine position within a few meters. Now we have weapons that are guided to their targets very accurately by using GPS to update their inertial systems. (This is discussed in further in the next section.) Dr. Foster and others bet on the outcome. One is reminded of the line from the book and movie, *Field of Dreams*: "If you build it, they will come" ("they," in this case, being the operators and users).

JDAM and the CAG for Bomber Weapons

In 1992, I was asked by the Assistant Secretary of the Air Force, John Welch, to head a group charged with making recommendations about how best to enhance the capabilities of the Air Force's fleet of attack aircraft (including heavy bombers) for operations with conventional (non-nuclear) weapons. The Cold War had just ended. With the demise of the Soviet Union, the requirement to deter aggression by posing credible threats of nuclear attack was greatly diminished in U.S. defense strategy. B-52 bombers had made an important contribution to the recent U.S. victory in the Gulf War, principally by bombing, day and night, Iraqi Army and Republican Guard forces in the field. But, with the exception of a small number of guided missiles, the B-52 was limited to dropping unguided weapons, which were ineffective against many types of targets and carried the risk of heavy collateral damage in many applications. The more-modern B-1 and B-2 bombers had played no role whatsoever in the Gulf War.

The Gulf War had highlighted the value of precision-guided conventional munitions for F-111s and F-15Es attacking both "strategic" targets, such as leadership facilities and infrastructure targets, and "tactical" targets, such as tanks in "tank plinking." Gen George Lee Butler, the commander of SAC (later STRATCOM), was determined to enhance the contribution of the heavy bomber force to future conflicts. Gen Mike Loh, then commander of TAC (later Air Combat Command), agreed with General Butler on the importance of getting the Air Force attack aircraft and the bomber fleet into the business of precision delivery of conventional weapons.

I formed a CAG, which I chaired, in order to explore the options for this. Other members of the CAG included a Colonel Richards from the planning division of SAC, a representative of the Armament Division at Eglin AFB, Russ Shaver from RAND, and others.

From the start, I was convinced that the best way to proceed would be to identify the specific capabilities that were most important for the bombers to have and then to define and evaluate concepts of operation that could yield those capabilities. This was the approach I had taken throughout my career when addressing problems of this type, and it had always proven fruitful. As with so many other problems relating to the Air Force's capabilities, it wasn't just a technical problem or a purely tactical one. Accordingly, I recommended to Colonel Richards that we form a CAG.

I worked with the members of the group, notably Russ Shaver and others at RAND, to define and identify the primary operational tasks to which the bombers might contribute in future conventional operations. We then defined concepts for accomplishing these tasks. New concepts came mainly from the development planner at the Air Force's Armament Division. We evaluated specific concepts according to the following criteria:

- technical feasibility
- operational effectiveness

- tactical and operational viability
- fiscal affordability.

After many meetings and in-depth discussions, the CAG reached a consensus on a set of measures that, if adopted, would greatly enhance the effectiveness of the bomber force in conventional military operations. Our main findings and recommendations were that all three bomber types should be equipped with a new system that would guide unitary weapons using signals from the GPS to update an onboard inertial system. This would involve fitting a GPS antenna on each aircraft and adding avionics and wiring so that the aircrew could program the weapons in the bomb bay using information on the aircraft's location at the time of a weapon's release. Kits would have to be developed that included inertial platforms, GPS receivers, fins, and servos to guide the weapons to their targets. These kits would be fitted to ordinary bombs, such as the Mark 84 2,000-pound unitary bomb.

Not long after developing our findings and recommendations, we briefed General Butler and, later, General Loh of TAC. Their reaction was quite favorable. They were especially enthusiastic about our main recommendation, namely, that all three bomber types be equipped with what would later become known as the Joint Direct Attack Munition (JDAM). It offered a way to provide the bombers with a near-precision weapon capability at a fairly modest cost. JDAM also had an advantage over such concepts as laser or electro-optical guidance in that its performance would not be affected by weather.

At this point, the work of the CAG shifted from conceiving and evaluating new concepts to determining how to get our primary recommendations, particularly the one regarding JDAM, implemented as quickly as possible. It soon became clear that we faced two major hurdles: the SPO for the B-2 and the so-called requirements generation system in the Pentagon. We devised strategies to overcome both of these.

In the Air Force, the SPO has the lead responsibility over the development of any new system. It manages the contracts for developing and building hardware and oversees the performance of contractors, who work for Air Force Materiel Command. Because the B-2 was still in development at the time that the CAG was working, its SPO, not the user command, had authority over any proposed modifications to the aircraft.

The head of the B-2 SPO had gotten wind of our work and was not enthusiastic about it. His objections to our recommendation that the B-2 be fitted with the capability to deliver GPS-guided ordnance sprang in part from a lack of a sense of urgency about our basic purpose and from his preoccupation with programmatic (as opposed to operational) issues. First, he felt that the B-2's primary purpose should remain what it was at the outset of the program—to be a nuclear bomber that could penetrate the heavily defended airspace of the (former) Soviet Union. The disappearance of the Soviet Union and the vastly reduced priority that U.S. policymakers were placing on nuclear deterrence escaped him.

Perhaps more important, the SPO was working feverishly to deliver the product within the time and budget constraints that the Air Force; DoD; and, most especially, Congress had set. Skeptical of the need for a state-of-the-art heavy bomber, some in Congress had moved to terminate the program. As part of a compromise to keep it alive, DoD agreed to a congressionally mandated cap on the overall cost of the program. By the time the cap was imposed, of course, most of the money allotted for the program had already been spent. So, to be able to put rubber on the ramp, the SPO considered it necessary to impose harsh discipline on both the contractor and on the Air Force to keep costs under control. For the Air Force, this meant "no new 'requirements' for the B-2."

Obviously, adding a precision conventional capability was a new requirement and was likely to raise the cost of the aircraft. The head of the SPO's first reaction was to "just say no" to the idea. When he saw that this would probably fail, he looked for the cheapest way to equip the B-2 with a precision conventional weapon. His answer: The B-2 would carry eight laser-guided bombs of the type already in the Air Force inventory. What made this a less-than-thoughtful solution was that, in the concept proposed by the SPO, some other entity would have to shine the laser beam that would guide the weapon to the target because integrating the laser designator on the B-2 would be costly.

In short, this state-of-the-art bomber, with its long-range and stealthy design, was designed to go where no other aircraft could go, but once it got there, it would need some other means of designating the target with a laser beam to enable it to attack the target with precision. Not surprisingly, we found the SPO's proposal less than compelling.

We reacted by mobilizing the operators against the SPO. It was an easy sell.

I got to work. First, I met with a former colleague who worked for Dr. Don Rice, then Secretary of the Air Force. Together, we made sure that the secretary understood both the importance of equipping the bomber fleet so that it could deliver modern conventional weapons and the inanity of the SPO's proposed solution. We then arranged for General Butler to call the secretary to communicate his views on the problems with the concept proposed by the SPO. That phone call was followed by a letter (which I drafted), signed by General Butler, outlining his recommended actions. The letter pointedly added that matters regarding the operational capabilities of major platforms, such as the B-2, were the proper purview of the operators and not the SPO.

The next task we faced was to start the program to develop and acquire JDAM weapons and to equip the aircraft with the necessary hardware and software to deliver them accurately. This meant getting the program through the first wickets in the acquisition process. "A Framework for Modernizing" (pp. 106–108) outlines some of the roadblocks that the acquisition process throws in the way of innovation within DoD. They were all present in spades in the case of JDAM. We were told that the first step would be to generate an MNS and to get it approved. We were also told that this process usually took 12 to 18 months. I found this to be an absurd requirement. The primary users of the capability—Generals Butler and Loh—had already identified the need for the bombers to have improved capabilities for delivering conventional weapons. What further "mission need" was needed?

Both generals (particularly General Butler) argued that the formalities imposed by DoDD 5000.1 were unnecessary and pernicious.[8]

---

[8] The current version of this document is Department of Defense Directive, 5000.1, *The Defense Acquisition System*, May 12, 2003.

Unfortunately, this time, their intervention had the opposite of the intended effect. Before he sent the letter I had drafted for him making known his recommendation to proceed with the program, General Butler added a paragraph of his own stating that a precision delivery capability for the B-2 and the other bombers was urgently needed and that, as a combatant commander, he should be able to attest to the "mission need" for the capability, thus obviating the need for a time-consuming preparation of an MNS in Washington. This last blivet stuck in the craw of the "requirements people" on the Air Staff and had the effect of making them dig in their heels. The result was that they were able to impose the requirement for a formal MNS, which caused a delay of more than a year in getting the program started.

Nevertheless, JDAM was operational on the B-2 by the time of Operation Allied Force in 1999, and it made a very significant contribution to the success of that operation. Not long after that, all the bombers were equipped to deliver JDAM. In 2001, JDAM-equipped B-1s, B-2s, and B-52s made major contributions to the success of Operation Enduring Freedom in Afghanistan, orbiting over the battlefield for hours at a time and delivering accurate weapons when called upon

---

Box 5.4
The Bottom Line(s)

If you want to spur innovation, empower a group to explore new operational concepts and give them one or more discrete operational problems to address.

Simply defining concepts that meet operational needs does not guarantee success. Sometimes the most important contributions one can make to innovation are to push new concepts through the bureaucracy.

The best way to generate high-level support for new concepts is to sell the user on them. Go to the top and show the four-star how you can solve a problem of importance to him.

Have the four-star (or in this case, two of them) sign a very short letter that contains a simple declarative statement to proceed to implement the concept.

in support of operations against the Taliban. As we foresaw, in scenarios in which access to bases close to the fight is problematical, heavy bombers equipped with precision weapons can be critically important to the combatant commander.

CHAPTER SIX

Analytical Tools

---

*Over the course of his career, General Kent developed and applied innu-
merable analytical tools and techniques. Many of these have already been
described in connection with the issues to which they were most relevant.
This chapter relates stories about analytical tools and techniques General
Kent used that were relevant to multiple issues. These include some of the
earliest combat models to run on computers. The most significant of these—
SABER GRAND—was developed under General Kent's guidance by Air
Force Studies and Analysis in the late 1960s. This model, which incorporated optimization algorithms, helped generate insights about investment
and employment options for a major conventional conflict in Europe. In
this chapter, General Kent also offers his views on how computer simulations should be used today in defense planning.*

Assessing the Effectiveness of Bomber Attacks

In the late 1950s, when I was head of the Weapons Division on the Air
Staff, we focused intently on the question of to what extent our bombers (mainly B-47s) could penetrate Soviet air defenses in a retaliatory
attack (after a Soviet attack on the United States). Later, the mainstay
of this retaliatory attack was ballistic missiles—SLBMs or ICBMs—
but the burden was on bombers in those days.

Gen Glen Martin, the deputy director of plans, revisited the issue
surrounding bombers over and over. He wanted the answer to many
"what-ifs": What if a certain fraction of our bombers was not launched
in time for "safe escape"? What if only a certain number penetrated the

outer zone (corridor) of Soviet air defenses? (He asked the same question for each zone; there were four in all.) What is the merit of attacking more targets (DGZs) that are "shallow" versus attacking higher-value targets that are "deep," by which we incur more losses of bombers before they reach their targets (release points)? And so on, and so on.

General Martin had directed these questions to a group of analysts in another division. He was not satisfied with the results of their efforts and informed me that, from now on, I would be in charge of answering his questions and would have oversight of this group for this purpose. General Martin inundated us with questions. He was asking questions much faster than we could generate answers. The "turn time" for answering each question was typically days—if not many days. The team worked long hours and on weekends. But the harder it worked, the more behind it was. Answers to questions spawned more questions.

Something had to give: Either we had to reduce General Martin's appetite for information, or we had to come up with an easier and faster way to generate answers. Knowing General Martin, the latter approach seemed more tractable. We decided to turn to what we called in those days a large-scale computer. The approach was as follows:

1. There will be a card (a punch card) for each individual nuclear weapon in the SIOP.
2. This card names the weapon (i.e., number so and so).
3. This card also names the delivery platform that carries it (i.e., what particular bomber, ICBM, or SLBM).
4. It indicates where the delivery platform is based.
5. It specifies the DGZ the weapon is to attack.
6. It designates the corridors (zones) of defense that the carrier has to penetrate to reach the stated DGZ.

Then, we would use Monte Carlo simulations to determine

1. which weapons (by tail number) were safely launched from their home bases
2. which weapons (by tail number) actually penetrated the defense corridors they were required to penetrate to reach their DGZs
3. the actual ground zero (AGZ) of each weapon.

We compared these AGZs to listings in a database that catalogued the locations of people, MVA, and industrial facilities and military facilities in the Soviet Union. We then could announce that, given a weapon of this yield detonating at this particular location, so many people would be killed and so many industrial (or military) facilities would be destroyed.

We needed help to do the programming. Computers were, in those days, not user-friendly. There were only a few people who were versed in the art of programming. We let a contract to IBM to provide four programmers. IBM was eager to help and welcomed the opportunity to show how its magic machines could be used to inform important decisions.

In about four weeks, we were up and running. I had convinced General Martin to cease and desist with his questions until we could develop this new tool. Fortunately, we did not have to start from scratch as far as the database was concerned. Now, the "turn time" was three hours, maybe less. Since the model was stochastic, we were obliged to make several trials for each "what-if." We settled on ten trials and then printed the results for the "median" trial. The printout was in two different formats: one for the generals and a more-complicated one for analysts. To be sure that we could stay ahead of the game of questions and answers, I gradually revealed that the turn time was now a matter of hours.

General Martin was very impressed with this effort. In fact, he made known our approach to the planners at SAC, and I lost the services of two of the experts from IBM.

To repeat, given a different input as to the probability of penetrating a stated defense zone, the computer sorted the "cards." In this "sorting," some weapons were taken out of the game. For those that penetrated to the release point, we used Monte Carlo to determine the AGZs. Incidentally, our high-speed computers consisted of tapes and wheels in a console taller than me, and it took awhile for each trial. Now, a computer, once programmed, could run each trial in the blink of an eye.

Assessing the Effectiveness of ABM Deployments

My first assignment after my year at the Center for International Affairs at Harvard was as Military Assistant to the Director of Research and Engineering. This assignment was about as "good as you get." My boss, Dr. Harold Brown, was a man of considerable intellect. The Secretary of Defense, Robert McNamara, relied heavily on the advice of Dr. Brown on an array of issues—many beyond the direct purview of DDR&E.

As it turned out, Dr. Brown, in turn, listened to others whom he had learned to trust. About a month after I arrived, the Army brought in a milestone study about the effectiveness of the Nike-X in terms of limiting damage to the United States from a Soviet attack with nuclear-armed ICBMs. The measure of merit used in this study was the proportion of the U.S. population surviving a stated Soviet attack, with and without a stated deployment of Nike-X interceptors.

The Nike-X interceptor was nuclear armed. It was "unguided" after launch and was launched to a point in space according to a tracking radar at the site. The P_k given an engagement was calculated as 0.50, hopefully greater. The range of the interceptor was stated as 8 nmi, so an interceptor at a site could defend an area with a radius of 8 miles.

The Army analysis was according to the following construct:

1. They had obtained "from the JCS" a copy of the RSIOP—a Soviet version of their attack on the United States, like the U.S. SIOP.
2. The Soviets would attack each DGZ with two missiles.
3. Each Soviet missile would employ one warhead and nine decoys.
4. Our radar (the Nike-X radar) could not distinguish between decoys and RVs.
5. The Nike-X battery defending the DGZ would fire two interceptors at each object; so, 20 Nike-X missiles would be fired for every Soviet ICBM that was engaged. Each Nike site had 40 missiles ready to fire, so each site could engage two Soviet ICBMs.

6.   Given the above, the Army then calculated the fatalities from the Soviet attack—with and without the Nike-X deployment of 40 interceptors per site.

The reduction in fatalities afforded by the Nike-X deployment was shown to be significant. I was getting bad vibes about the analysis from the start. I remembered the advice of Dr. Schelling, who taught "The Strategy of Conflict" in a course at Harvard. He cautioned us to think carefully about which side has the "last move." The construct the Army used implied that the United States had the last move: The Soviets defined their attack; that is, they chose their DGZs and assigned two missiles to each one. The United States was made aware of the character of the attack and responded accordingly, by deploying 40 interceptors at each DGZ.

The reality, as I saw it, was quite different: The Soviets would observe the deployment of the Nike-X battery by battery and tailor their attack accordingly. They were not obliged to attack the defended areas defined by our deployment. Rather, they could attack undefended areas if they chose to do so. Surely, then, the Soviets had the last move. And if so, the difference in U.S. fatalities with and without the Nike-X would surely be much less than shown by the Army analysis.

I was the new man on the block, so I held my tongue while the Army was briefing Dr. Brown. However, when the briefing was over, I followed Dr. Eugene Fubini back to this office. Dr. Fubini was Dr. Brown's principal deputy and a man with a razor-sharp mind. I had hardly finished explaining the "last move" problem when Dr. Fubini stood up abruptly: "Of course, of course," he said. "Come with me to Dr. Brown's office." There, the argument was repeated. "You are absolutely correct," Dr. Brown said. "Call the Army back here."

Dr. Brown explained his concern to the Army. "We cannot use this analysis as a basis for informing our decision about deploying Nike-X," he stated. This demanded a new question: How *do* you go about deciding where and how many interceptors will be deployed when the Soviets have the last move? The Army and its contractor, California's Stanford Research Institute, were back to the drawing board.

A day later, the other shoe fell. Dr. Brown directed me to "think through this problem." Ugh.

After wrestling with the problem for a week or so, I had made some progress, but not much. I now realized there was a related issue: the firing doctrine at the battery. How many interceptors do you employ per "object" in the attack if you do not know how many objects are yet to appear? You must balance "leakage" against "exhaustion." From the standpoint of leakage, you employ many interceptors per object. But the more you employ, the sooner your supply of interceptors is exhausted. The analysis by the Army had no problem in this regard. The Soviets always fired two missiles per DGZ, and knowing this, we always engaged all 20 of the objects. But if the Soviets attacked some of the DGZs with three missiles instead of two, the defenders would go from modest success to certain failure. If the attack is two RVs per DGZ, there is a 0.56 probability that no RV (warhead) will detonate[1]; if three, there is an absolute certainty that one will because the defender has no more missiles with which to engage the third RV.

Then luck came my way. Someone told me that two analysts at Bell Telephone Laboratories had issued a paper on this very subject. They were Dr. Robert Prim and Dr. Thornton Read. I called Dr. Prim. He was delighted to come to the Pentagon and show me their work.

Their construct was absolutely elegant. The primary elements were as follows:

1. The defense charges a price, p, for a stated target.
2. The price charged, p, is directly proportional to the "worth," W, of the target.
3. A defense is defined by the ratio of W to p. They called this ratio lambda, λ.
4. If the p of a target (a DGZ) is one, no interceptors are required.
5. If p is two, then one interceptor against the first RV is required. If the P_k of the interceptor is 0.50, the Soviets gain one-half of

---

[1] This is so because $0.5^2 = 0.25$ and $0.75^2 = 0.56$.

the worth of the target with the first RV and 1.0 × one-half with the second RV.

Absolutely elegant.

I now turned to the task of calculating the "effectiveness" of a stated Nike-X deployment based on the Prim-Read theory. Given that the P_k given an engagement was a certain value and setting aside the issue of decoys for the moment, I devised an expression that gave the number of interceptors required as a function of (1) the price you charge and (2) the P_k given an engagement. In it,

$$I = \frac{\ln(p!)}{-\ln(1-P_k)},$$

where I is the number of interceptors. For $p = 4$ and $P_k = 0.5$, this yields a total of 4.59 interceptors to charge a price of four RVs. If the price is four RVs,

1. There is a one in four probability of penetration for the first RV; two interceptors are required.
2. There is a one in three probability of penetration for the second RV; 1.59 interceptors are required.
3. There is a one in two probability of penetration for the third RV; one interceptor is required.
4. There is a 1.0 probability of penetration for the fourth RV; 0 interceptors are required.

The total is 4.59 interceptors, just as the formula said. You will note that the "expected return" for each RV is the same:

$$\text{RV}_1: \quad \frac{1}{4} \times 1.0 = \frac{1}{4}$$

$$\text{RV}_2: \quad \frac{1}{3} \times \frac{3}{4} = \frac{1}{4}$$

$$RV_3: \quad \frac{1}{2} \times \frac{2}{4} = \frac{1}{4}$$

$$RV_4: \quad 1.0 \times \frac{1}{4} = \frac{1}{4}.$$

To make it easier to do "what-ifs," I used the formula to plot the number of interceptors required as a function of price charged for a range of P_k values (see Figure 6.1 for an example). With this "analog computer," calculating the number of interceptors required became a simple matter of reading values off of a graph.

Once you understand the construct, the rest is easy:

1. Postulate λ, the ratio of W to p. The smaller the value chosen for λ, the larger the price you must charge for a target of a given worth.
2. Decide how many DGZ areas you intend to defend.
3. Calculate how many interceptors are required for each DGZ. This number depends on the λ you have chosen, the worth of that particular DGZ, the P_k of the interceptor, and the number of objects (RVs and decoys) in each missile. In the example above, since each Soviet missile carried ten objects (one RV and nine decoys) and the Nike-X radar could not distinguish between RVs and decoys, the total number of interceptors required to charge a price of four is $4.59 \times 10 = 45.9$.

Now make a plot of the worth destroyed as a function of the number of Soviet missiles in the attack. For this, I put the number of Soviet missiles on the abscissa and "worth destroyed" for the defended areas on the ordinate. This is a straight line starting at the origin and at a slope of λ. The line ends at the total worth contained in the "defended areas." If the Soviet attack includes the "undefended areas," tack on the worth destroyed per missile expended in the undefended areas.

If you have defended all "areas" whose worth is equal to or greater than λ, the marginal return of worth for each Soviet missile expended

Figure 6.1
Interceptors Required Versus Price Charged for a Particular Target

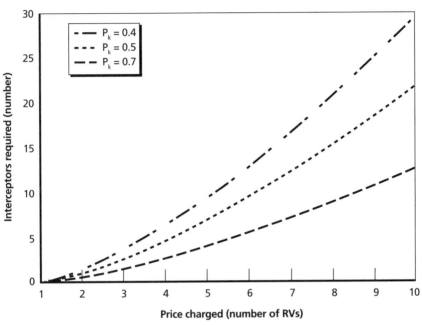

is exactly λ at the juncture of the lines of defended and undefended areas.

An example of such a plot is shown in Figure 6.2. The figure reflects calculations of worth destroyed as a function of the size of a Soviet attack for two cases: (1) no defense and (2) a defense that allocates interceptors according to the logic of Prim-Read. Both cases in this somewhat notional example assume a target set containing 1,600 DGZ areas and a total population of 200 million. The distribution of the total population among the 1,600 DGZ areas is assumed to obey a Pareto distribution with an exponent of one-half. That is,

$$W_{cum} = 200 \left(\frac{n}{1,600} \right)^{\frac{1}{2}}$$

Figure 6.2
Comparison of Worth Destroyed for Two Cases

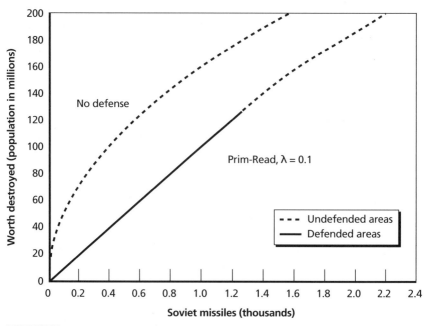

RAND OP223-6.2

or

$$W_{cum} = 5 \times \sqrt{n} \, ,$$

where W_{cum} is the cumulative worth and n is the number of targets over which "worth" has been accumulated. In the no-defense case, this relationship is also what defines *worth destroyed* as Soviet missiles are traded one-for-one with DGZ areas, starting with the most lucrative areas to maximize the "expected return" of each missile.

In the case of the Prim-Read defense, λ has been set at 0.1. Thus, we want to defend DGZ areas up to the point where the "expected return" of Soviet missiles is 0.1 million per missile. The next issue that arises is how many targets we should defend. This question can be addressed target by target, but I find it convenient to work with sets of targets instead. Since cumulative worth obeys a Pareto distribution with an exponent of one-half, we can define 40 sets (40 being the square root of 1,600) with the following properties:

1. The number of targets in the kth set is equal to $2k - 1$. The first set contains the most valuable target; the second set contains the next three most valuable targets; the third set contains the next five most valuable targets; and so on.
2. The number of targets through the kth set is equal to k^2. There are four targets in the first two sets, nine targets in the first three sets, 16 targets in the first four sets, and so on.
3. All sets are equal in value. Although each set contains a different number of targets, the value of the first set

$$5 \times \sqrt{1} = 5$$

is the same as the value of the 40th set

$$5 \times \sqrt{40^2} - 5 \times \sqrt{39^2} = 5$$

and every set in between. This is simply the total worth of the target set divided by the number of sets, which is the square root of the total number of targets. In this example, each set contains a worth of 5 million.

Since $\lambda = 0.1$ and since each set has a worth of 5 million, the defense should charge a price of 50 for each set to be defended. If there are more than 50 targets in a set, it does not need to be defended, since the expected return per missile in that set would be less than 0.1. The 25th set contains 49 targets ($2 \times 25 - 1 = 49$). The 26th set contains 51 targets ($2 \times 26 - 1 = 51$). The expected return per missile expended in the 25th set is slightly greater than 0.1, and it is slightly less than 0.1 in the 26th set. Thus, we defend 25 sets containing 625 targets with a population of 125 million. The remaining 15 sets (975 targets with a population of 75 million) are left undefended.

In the absence of a defense, the Soviets could destroy half the total worth of the target set by attacking the 400 most lucrative areas with 400 missiles. In the presence of a Prim-Read defense with $\lambda = 0.1$, the Soviets would need 1,000 missiles to inflict the same level of damage. Moreover, only defended areas would be targeted, since attacking these DGZs brings the highest "expected returns" for Soviet RVs and the attack is not large enough to expect to destroy the total worth in the defended areas (a population of 125 million). Undefended areas would be left alone unless the Soviet attacks involved more than 1,250 missiles.

It is worth noting that, while the defense significantly raises the price of achieving a specified level of damage, it is not an inexpensive proposition for the defender. For example, the third set contains five targets with a population of 5 million. The population of an average target within the set is 1 million. Since $\lambda = 0.1$, the defender should charge a price of 10 for the average target, which would require 21.8 interceptors for the average target. Thus, roughly 109 interceptors are required to defend the set of five targets if there are no decoys or if the radar can distinguish between RVs and decoys. If we assume, as before, that each Soviet missile in the attack carries one warhead and nine decoys and that the Nike-X radar cannot discriminate between RVs and decoys, roughly 1,090 interceptors would be required to defend the five DGZ areas in the set. Around 11,000 interceptors would be required to cover the 625 defended DGZ areas.

Dr. Brown was quite impressed. He directed that the Army use this construct. They accepted the construct, but it took the Army's contractor a while before he could get the computer to do his bidding. Accordingly, I spent a good deal of time doing "what-ifs" using a spreadsheet I had developed and a Friden calculator.

Doing "what-if" calculations on a target-by-target basis would have been extremely time consuming. After wrestling with the problem for a bit, I realized that I could reduce my workload substantially if I used a Pareto distribution with an exponent of one-half to represent the distribution of worth among U.S. targets. I could then take advantage of the properties of the distribution to define equal-value sets of targets. I placed the most valuable target in the first set, the next three most valuable targets in the second set, the next five most valuable targets in the third set, and so on. The total number of sets was equal to the square root of the total number of targets. For each set I calculated the number of interceptors required to charge the desired price for an average target and then multiplied by the number of targets in the set to determine the total number of interceptors required to defend the set. The set-based approach dramatically reduced the number of "rows" required on a spreadsheet. It was far easier and much faster to have to work only with, for example, 40 sets instead of 1,600 individual targets. My "turn time" was about two or three hours.

With a high-speed computer to run the spreadsheet, the turn time could have been reduced to minutes. Accordingly, Dr. Brown directed the Army to instruct its contractor to do the "what-ifs" by running my spreadsheet with a high-speed computer. Following several unsuccessful attempts by the Army's contractor, Dr. Brown turned to the Institute for Defense Analyses (IDA) to do the "what-ifs." IDA's project leader then made a serious mistake. Instead of running the spreadsheet, he chose to render a critique of the whole calculus. He let a contract to this end to a professor of mathematics at New York University. This professor's finding: While Colonel Kent's methodology may be appealing on some counts, it has no firm foundation mathematically.

The professor offered a construct that had two features: (1) It would not work in terms of deriving the relationship among size of attack, interceptors deployed, and worth destroyed, and (2) it gave wrong answers for items you could calculate. He had made a careless error in signs: He had a plus when it should have been a minus.

Both Bob Prim and Thornton Read were infuriated at this report. After all, it was their construct that the professor had criticized. Dr. Prim went directly to Dr. Brown. IDA was again directed to do the "what-ifs," but with no changes as to the spreadsheets.

The Prim-Read theory still stands as an exceptional piece of work. Analysts over the years have amended the calculus so as to account for noninteger solutions. (The defense cannot fire 1.59 interceptors at an object.) But the basic construct still stands. It was my luck to come across it and use if to good effect.

And so it was that my involvement in the calculus of the contribution of active defense led to thinking about the contribution of other means of limiting damage to the United States from attack by nuclear-armed ICBMs: (1) counterforce operations, (2) active defense, and (3) passive measures. This led to the defining study, "Limiting Damage to the United States," pp. 43–50.

Providing Insights with the SABER GRAND Model

One effort I worked on while at AFSA that deserves special attention was centered on the SABER GRAND model. The primary purpose of the model was to have a tool that could be used to shed light on the value of various investments (for example, whether to buy more aircraft of a given type or to buy better weapons, and which kinds; whether to harden existing airbases or to invest in something else).

The time was the late 1960s and, not surprisingly, the scenario we used was the Central Front in Europe. The analysts were to develop a computer model that captured the utility of allocating allied aircraft to the following four missions:

- Attack the Soviet and Warsaw Pact airbases to destroy their combat aircraft and otherwise curtail their capability to generate sorties.
- Defend our own airbases against attack by Soviet combat aircraft.
- Defend against Soviet air attacks against our troops on the battle-field, including sweeps to engage Soviet combat aircraft.
- Attack enemy ground forces, both through interdiction and close air support.

Our measure of merit was as follows: Blue ordnance delivered against Red troops minus Red ordnance delivered against Blue troops. The model produced day-by-day accounts of the following:

- Red (Warsaw Pact) aircraft remaining, by type
- Blue (NATO) aircraft remaining, by type
- the ordnance delivered by Blue aircraft against Red troops
- the ordnance delivered by Red aircraft against Blue troops.

We did not model the interaction of Blue and Red ground forces. Our focus was on how best to invest resources in air assets and how best to employ them. We did take into account the contribution of air forces to the joint campaign, since our primary measure of merit was Blue ordnance on Red troops minus Red ordnance on Blue troops. The Blue combatant commander wants all he can get of Blue on Red. At the same time, he wants to minimize Red on Blue. (Unlike many other theater-level combat models, we did not seek to estimate the movement of the forward line of troops.)

We had a plot of Blue on Red on the ordinate and Red on Blue on the abscissa (see Figure 6.3). Straight lines at a 45-degree slope represent lines of constant difference. The time horizon was typically a 10- to 15-day campaign. This was important: If the campaign lasts only one day, the Blue commander would naturally allocate heavily toward attacking Red troops. But if the campaign lasts 10 days, there would be

Figure 6.3
Measures of Merit

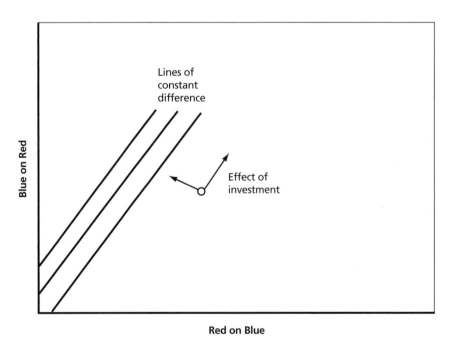

a large payoff in destroying enemy combat aircraft in the first few days: These aircraft would not be able to kill Blue troops or destroy Blue aircraft in the remaining days.

As we continued to develop the model, we asked some senior people on the Army staff to review our assumptions. Specifically, we asked for their judgment regarding the slope of the lines trading off Blue ordnance on Red versus Red ordnance on Blue. Specifically, we thought that perhaps some Army officers would place more value on preventing Red on Blue over inflicting Blue on Red. About as many Army people voted for a steeper line as voted for a more shallow one. So we kept the slope at 45 degrees, which implied equal weight.

After a great deal of work, we had a model up and running. Four young officers did the work: Maj Leon Goodson, Capt Scott Myer, Capt Lou Finch, and Lt Ken Hinkel. In the first cut at this model, a Blue commander defines the allocation of Blue sorties to the four different missions—in effect building an air tasking order for each day of the campaign; the same goes for Red.

We soon discovered what should have been obvious: For the same quality of forces (Blue and Red), the outcome varied markedly depending on the allocation choices made by the "commander" on each side. A particular Blue commander might do very well against a certain Red commander. But if we switched commanders, the Red force was dominant. In fact, the brilliance (or ineptness) of commanders accounted for such great differences in the outcome that the marginal return of a stated investment was lost in the noise. Obviously, this fact limited the utility of the model as a means for determining the merit of different investment packages.

Accordingly, I put the group of four back to work, this time with the task of developing an algorithm whereby the computer would determine the best allocation, day by day, that would maximize the measure of merit for a stated length of campaign. In about a year, they finally had developed such an algorithm. We satisfied ourselves that they had the algorithm about right. No human operator setting the allocation could "beat" the computer. We had taught the computer how to "play chess," and now the computer could beat the experts. The only operators who could come close to the computer were the four officers who had studied the computer's allocations, day by day, and did their best to copy them.

There is a lesson here. Developing the algorithm was a long trek. In fact, more than once, the four analysts working on it wanted to abandon their efforts because they saw little hope of succeeding. Fortunately, I enjoyed the luxury of knowing that there was no great downside if we failed. I had kept our effort under cover. Since there was no great cost associated with failure, we would keep on trying. "How long?" the team would ask. "No set time," I'd reply. "You will continue until you succeed. The only way you get into my doghouse is if you quit trying." They redoubled their efforts.

Finally, the day came when we revealed this magic model to the Chief of Staff. He was quite impressed. The next day, I received a call from General Momyer, the commander of TAC. He had just arrived back at TAC from a visit to the Pentagon. General Ryan, the chief, had told General Momyer of our effort. "I thought you would have run it by me first," Momyer insisted. Gulp. We were getting off on the wrong foot with the commander of TAC.

"When do you want to see it?" I offered.

"Tomorrow morning at nine. My office."

"Yes, sir!"

For better or worse, I could not make the trip due to a previous engagement that I could not break. However, I did have Lt Col Larry Welch go with the four young officers. They needed some sort of cover, and Colonel Welch had served under General Momyer in Vietnam. They went to Langley and presented their work.

Colonel Welch reported to me the next day an account of the meeting. Things were tense at first, but the general, at the end, seemed impressed by the effort. The next day General Momyer called me again. "I think we have something very useful here. I want to see it again."

"When?" I asked.

"Tomorrow morning at nine. My office."

"Yes, sir."

At the second meeting the general stated how he intended to use the model. His interest in the model sprang not so much from a desire to explore the relative utility of alternative modernization initiatives, but rather from a desire to educate people about the proper employment of airpower. Specifically, General Momyer wanted to use the model (1) to train future officers in the art of running a campaign and (2) to show his fellow Army officers that gaining control of the air at the outset of a campaign would allow airpower to do more to support the ground campaign. General Momyer had long espoused this doctrine. Now he had analytic support for it, since the optimum allocation in our runs was heavy on attacking enemy airfields and gaining control of the airspace in the opening days.

General Momyer said that he wanted us to hand the model to TAC and train his people to run it. I offered instead that we would do the runs he wanted and that my people would do whatever briefings he directed. He accepted this offer, and for some time I saw little of Major Goodson. He spent much of his time over the coming months at Langley. Major Goodson was sent by General Momyer to brief the Military Committee at NATO headquarters, followed by a tour of major command headquarters throughout Europe.

General Ryan wanted all important investment options evaluated using SABER GRAND. General Momyer wanted operators to learn from it how to run an air campaign. It was also used by the chief in briefings to leaders of the Army, to show that the Air Force was not fighting for dominance of operations in the air for its own sake but rather as a critical prerequisite to supporting operations on the ground.

There is nothing comparable to SABER GRAND today. That is, we do not have a model built around fairly straightforward metrics that assesses the utility of the air component of a joint campaign. One reason for this is that there is no longer a single, set scenario like the Central Front to use as a yardstick against which to assess the utility of various forces. The other reason for the demise of SABER GRAND is that Major Goodson, Captain Myer, Captain Finch, and Lieutenant Hinkel all left AFSA around the same time. I had left AFSA in 1972, around the time that General Ryan retired and General Meyer, the Vice Chief of Staff, moved on to command SAC. The people who remained behind in AFSA had less experience with the model, and it fell into disuse.

In the late 1980s, RAND developed a model called TAC SAGE that was somewhat like SABER GRAND. It allowed researchers to explore questions relating to the optimum allocation of air assets in conflicts on the Central Front. But, for one reason or another, it never assumed the prominence afforded SABRE GRAND, the original effort.

Another Episode with SABER GRAND

In the early 1970s, questions arose about the relative strengths of the NATO and Warsaw Pact air forces on the Central Front. Congress mandated that DoD conduct a joint study of the issue. The Secretary of Defense, Melvin Laird, directed Dr. Enthoven to conduct this study—with Air Force participation. I was not happy with this setup because it placed PA&E in the lead, leaving us at a disadvantage.

In due time, Dr. Enthoven convened a meeting at which he set forth the construct of the study:

- The measure of merit was to be the "combat potential" delivered by Blue (NATO) over a period of 10 days, compared to the combat potential delivered by Red during the same period.
- The inputs would include the number, tons of ordnance carried, and sortie rate of each type of aircraft (Blue and Red).
- The range to which munitions were to be delivered was specified. Several different ranges would be examined.[2]

The "analysis" then became a simple spreadsheet affair. Once there was agreement as to the sortie rate and combat load, the rest was transparent and simple arithmetic.

I let it be known that we (the Air Force) would work with the study team to determine the numbers for sortie rates and combat loads and opined that we could probably reach agreement on these matters.

---

[2] Because Red aircraft were typically shorter-legged than Blue aircraft, making the distance longer weighed in favor of the Blue side. At the extreme, one could specify a distance so long that some Red aircraft had a combat load of zero, or near zero, at the range.

However, we could not agree to the overall construct of the analysis as presented. I pointed out that the Warsaw Pact had many more fighter aircraft than NATO and that, while these aircraft could not deliver a great deal of ordnance, they could be quite effective in blunting the attacks by NATO aircraft against their airfields and, as well, attacks by NATO aircraft against Red ground forces. The construct presented would not capture the value of either side's counterair operations. In the approach offered by PA&E, the opposing air forces did not fight each other. This, I stated, was a fatal flaw.

When opposing forces do not interact with each other, the trade-off between numbers and quality is according to the linear law. That is, the correlation of power (CP) is

$$CP = \frac{B_0}{R_0} \times \frac{\gamma}{\lambda},$$

where B_0 is the number of Blue aircraft, R_0 is the number of Red aircraft, γ represents the quality (effectiveness) of each Blue aircraft, and λ represents the quality of each Red aircraft.

On the other hand, if the forces interact with each other, the correlation of power is defined by

$$CP = \left(\frac{B_0}{R_0}\right)^2 \times \frac{\gamma}{\lambda}.$$

Here, any numerical advantage operates at the square. If Red has twice as many aircraft as Blue, and the linear law applies, then the CP is one if each Blue shooter is twice as good as each Red. However, if the square law applies, then given

$$\frac{B_0}{R_0} = \frac{1}{2}$$

and

$$\left(\frac{B_0}{R_0}\right)^2 = \frac{1}{4},$$

each Blue has to be four times as good as each Red for the *CP* to be one.

Obviously, then, the assumption one makes about the interaction (or lack thereof) between the two forces will have an important bearing on the outcome of the study. Preliminary calculations showed that, if PA&E's construct stood (and the linear law applied), NATO's air forces would be made to appear somewhat superior to those of the Warsaw Pact—an outcome that not only flew in the face of reality but that could be prejudicial to the Air Force's effort to develop and field more-effective conventional forces.

The history of World War II (and most other conflicts) shows, of course, that air forces of opposing sides do indeed interact. Because they do interact, the analysis is ever so much more complicated.

I told Dr. Enthoven that we had a model, SABER GRAND, which we could use to capture these interactions. Dr. Enthoven responded that he was quite familiar with Lanchester laws, as well as with SABER GRAND. He went on to state that he was not going to use our model in this study. "If you don't use SABER GRAND," I asked, "what model will you use to capture the effects of counterair interactions?" He made no reply.

And so the battle was joined: interactions or no interactions? I wrote a short letter to Dr. Enthoven and pointed out that, since his construct did not take into account counterair interactions, it was dead on arrival. I added that the effort should no longer be labeled a joint study because the Air Force did not concur with the fundamentals of his analytical construct. I instructed my staff to continue to participate in the study run by PA&E and to negotiate in good faith about such issues as sortie rates and combat loads. But under no circumstances were they to buy into the overall construct of the study or to agree at any time that this was a joint study.

I kept the chief and vice chief informed of these happenings and laid out the following plan of action:

- Reject PA&E's analytical construct, and conduct our own study in parallel to theirs.
- Agree with PA&E's assumptions about the sortie rates that Air Force aircraft could sustain in wartime. (PA&E maintained that it was possible for F-4 units, for example, to generate two sorties per day—a higher rate than we were then planning to mount.) This would require manning the squadrons with more combat crews and more maintenance people.
- Wait until PA&E finishes the report and distributes it to the Chief and the Secretary of the Air Force for comment. At this point, object strenuously to the overall analytical construct of the study, pointing out the absurdity of assuming, in effect, that there are no interactions in counterair operations between Red and Blue.
- Deliver our comments directly to Secretary Laird in a meeting with the chief.

The chief agreed to this plan of action. As we awaited PA&E's submission of the report, Vice Chief Gen J. C. Meyer wanted to know more about Lanchester's linear law and square law, so I tutored him a bit on these.

When PA&E submitted its report for comment, the leaders of the Air Force requested a meeting with Secretary Laird to discuss the matter. At that meeting, Dr. Enthoven presented his findings, emphasizing that the analysis showed that the Air Force should make the necessary investments to significantly increase the sortie rate of its units in Central Europe, especially F-4 units.

The chief delivered the Air Force's response. He noted in passing that the Air Force had no objection to the finding that sortie rates could and should be raised, pointing out that the requisite investments for this were already being made. Then he focused the discussion on one point: The Air Force, he said, rejected the finding that the "combat potential" on the Central Front favored NATO air forces. Our objection was based on the fact that the analysis by PA&E was fatally flawed because it did not consider the interactions of combat aircraft in counterair operations.

The vice chief then rendered, for Secretary Laird's benefit, a short dissertation on the linear law and the square law. The Warsaw Pact had more combat aircraft but of lower quality per aircraft. If the linear law applied, the NATO force was superior; if the square law applied, the Warsaw Pact force was superior.

Following this, Major Goodson showed the results of our analysis based on SABER GRAND. He pointed out that, if the model was run under the assumption that fighters on both sides had zero capability to kill other aircraft in counterair operations, the algorithm allocated no (zero) aircraft to these missions. Under these conditions, we gained outcomes in the upper right corner in the domain of "Red on Blue" (abscissa) and "Blue on Red" (ordinate). These results were close to those reported by PA&E, showing a balance of capability in favor of NATO.

However, if counterair interactions were modeled, our analysis yielded quite a different answer: The amount of ordnance delivered by both sides would be greatly reduced; most important, Blue's delivered ordnance is reduced more than Red. (The numerous Red fighters take their toll by killing Blue aircraft and by requiring Blue to devote more effort to counterair.) The measure of merit—"Blue ordnance delivered" minus "Red ordnance delivered"—is now a negative number: The correlation of power is now in favor of Red.

At this point, the vice chief delivered the final blow: It would, he observed, be awkward (to say the least) for DoD to send a report to Congress that purported to yield insights on the relative strength of the NATO and Warsaw Pact air forces if the analysis did not address the effects of the interactions of these combat aircraft in counterair operations. He observed dryly, "The PA&E report acts as if occasional, inadvertent midair collisions are the only interaction between the two forces."

"I will get back to you," said Secretary Laird, through tight lips.

That afternoon, we reconvened in the Secretary of Defense's office. This time, the vice chief and I were the only Air Force representatives present. The secretary was brief. He said that he could not use the PA&E report in its current form, yet a report was due to Congress in two days. Accordingly, we were directed to fashion a new report. It would be a joint report. We did so, and the secretary issued the new, "joint," report two days later under his name.

The new report incorporated the recommendations of the earlier report as to sortie rates. But it did indeed include the findings that the correlation of power was in favor of the Warsaw Pact air forces rather than the NATO air forces.

The question then arises: Why all this effort to reverse a finding in a report about the balance of air force capabilities on the Central Front? The answer is that we were concerned that an erroneous portrayal of that balance would undercut the advocacy for modernization efforts that we believed were essential to NATO's deterrent and defense posture. If NATO forces are superior, the argument might go, why appropriate money to modernize them? We therefore felt that this was an argument we had to win.

The long recital of this particular episode underscores several points:

- Pick the arguments you absolutely have to win.
- Maneuver the events and the process so that the top leaders of the Air Force are involved.
- Isolate and dwell on the one key issue or word (in the episode above, *interaction*).
- Have a well-thought-out plan of action.
- Do your homework.

If you follow these steps, surely you will prevail.

What we accomplished was quite a feat. The Secretary of Defense used the analysis by the Air Force and rejected the analysis by PA&E. This does not happen all that often.

One final note: I am not known as an enthusiastic proponent of campaign-level models. Too often, I have found that their opacity and complexity hide a variety of sins on the parts of both programmers and users. And placing a complex model at the center of an analytical effort tends to drive activities toward "getting the model to run," rather than developing knowledge and insights about the problem at hand. But for certain types of problems, a campaign-level tool is essential. We relied on SABER GRAND in part because, having built it at Studies and Analysis, we knew that it was fundamentally sound. Also, the model used a simple measure of merit: The difference between Blue ordnance dropped on Red and Red ordnance dropped on Blue. This allowed us to avoid the pitfalls and additional complexities associated with trying to ascertain movement of the forward line of troops—the more common measure of outcome. That said, my record with regard to the development of campaign models is not spotless, as the following section relates.

Fostering Campaign Models

In 1972, I was the head of WSEG. My bosses were ADM Thomas Moorer, Chairman of the JCS, and Dr. John Foster, Director of DDR&E. In my capacity as head of WSEG, I had persuaded Dr. Foster to allocate $3 million to develop campaign models to facilitate the evaluation of various investments to improve the operational capabilities of conventional forces.

I knew that developing rigorous models of anything as complex as large-scale conflict would be a daunting task. I recommended that we proceed in parallel with three efforts, allocating, at first, $1 million to each of them. Dr. Foster agreed, and in due time, three contracts were awarded. The winners were

1. IDA, under Dr. Jerry Bracken
2. Vector Research, under Dr. Seth Bonder
3. Lulejian and Associates, under Dr. Lulejian.

All three contractors developed models wherein a computer ran the campaign—more or less. There was some opportunity for human intervention along the way, but by and large, the computer ran the game. This differed from other models in which the operator runs the campaign and calls upon the computer to run particular calculations.

When the computer runs the game, it does so in accordance with a preset scenario defined by the operator. The computer does not amend the scenario in the presence of unfolding events. As such, these models do not do a good job of capturing what might be called the *operational art of war*, in which strategists on both sides adjust their campaigns dynamically in reaction to unfolding events.

In intervening years, I have come to believe that I should have framed the problem differently. In an ideal situation, our models would be developed and employed as follows:

1. The operator (commander) runs the overall campaign "hands on."
2. From time to time, the operator needs to know, in a predictive sense, the outcome of a certain event so that the operator has some insight as to whether to maneuver to cause the event or whether to maneuver to avoid the event.
3. The events include skirmishes, engagements, firefights, or even battles, whatever term you wish to use.
4. The outcome of an event depends on the "conditions":
   a. the correlation of forces (by type and number)
   b. the kill potential of each type of agent (forces consist of agents)
   c. the conditions created by "maneuver" (such as prepared defense, hasty defense, ambush, enemy on alert, enemy asleep)
   d. the conditions controlled by nature (such as night, day, fog, rain, snow, terrain)
   e. the conditions the commander can control (such as time, place).[3]

The "art of war" at the operational (maneuver) level, then, has a major role in determining the conditions under which each major event

---

[3] While certain conditions are controlled by nature, the commander has, within the constraints imposed, some control over important conditions. For example, in many cases the time at which to initiate an operation can be chosen so that other conditions are as favorable as possible.

(a firefight or a battle) in a campaign takes place. But it plays a similar part in determining the outcome of that particular event, given the conditions under which the event takes place. Thus, to adjudicate these events, the operator needs a quick and "not-so-dirty" means of determining outcomes of postulated events under postulated conditions.

Dictionaries define a *campaign* as a series of military operations for a common purpose. Other things being equal, if the Blue commander at the operational level maneuvers his forces so as to cause events to happen at the time and place of his choosing, he is apt to carry the day, as far as the battle is concerned. If he does well in successive battles, he is apt to win the campaign, and so on.

Take the example of the Revolutionary Army (Blue) defending the city of New York against the British and Hessians (Red) in 1776. By all accounts Blue was outmaneuvered, for one reason or another, by Red, and was badly defeated in a series of skirmishes—at Long Island, Staten Island, Friendship Heights, and elsewhere. The Blue commander, General George Washington, withdrew his forces to Valley Forge. At this point, the Americans' cause was in dire straits: American forces were demoralized, ill fed, ill clothed, ill equipped, and sparsely manned. At this juncture, General Washington took a huge gamble. He crossed the Delaware and engaged the Hessians at a time of his choosing, at night, and on Christmas Eve. The conditions, as he planned, were in his favor. Most important, Blue forces achieved surprise, and virtually all Red agents were ill prepared to engage the attackers. Blue carried the day in that event. Trenton was, in the greater scheme of things, not a major battle, but it provided a major boost to the morale of the independence movement at a crucial juncture in the war.

My purpose in relating this piece of history is to ask whether a campaign model, as we know it today, would have been of any help to General Washington in determining his plan of maneuver for the next several days. If an operator dictates a scenario to a computer, the concept of crossing the Delaware is not likely to appear. The general has to have a plausible theory of success in the event. This is why I favor models that adjudicate the outcomes of discrete events, as distinct from models that strive to simulate an entire campaign.

In retrospect, my focus in 1972 was one or two levels too high. The focus for the development of combat simulators should have been at the tactical-engagement level and not at the level of extended battles or campaigns.

The Trade-Offs Between Numbers, Yield, and CEP in Hard-Target Kill

From time to time (in fact, too often according to my tastes), the AFSC program office for ballistic missiles would present to the Air Staff proposals to modify and modernize the Minuteman missile. During my tenure in AFSA, the Ballistic Missile Office (BMO) often sought to increase the HTK capability of the Minuteman force. HTK was defined as the number of Soviet missile silos that could be attacked by the Minuteman force with a damage expectancy of 0.9.

HTK can be increased in the following ways:

- increasing the yield of the warhead, which entailed a newer design and the use of more enriched uranium in each warhead
- decreasing the CEP using a new and improved guidance system
- simply increasing the number of missiles deployed
- pursuing some combination of all three approaches.

I was generally skeptical of these proposals. In the first place, the argument for seeking more HTK was not compelling: Even if the United States were to launch a first strike, the relationship between increased HTK and significantly reduced damage to the United States was obscure. The second problem with the proposals had to do with the "packaging" of increased yield, improved accuracy, and greater numbers of warheads into a single proposal. BMO's presentations about the contribution of each component to the overall measure of merit were always opaque.

To gain insight into these matters, I developed the following construct:

- The measure of merit is the number of Soviet silos attacked at a stated DE of 0.9.

- The probability that a Soviet silo will survive an attack by one U.S. RV, P_s, is given by

$$P_s = 0.5^{\frac{LR^2}{C^2}},$$

where LR is the lethal radius of the warhead, and C is the CEP of the RV.
- Lethal radius is related to the yield of the warhead, y, by

$$LR = \lambda y^{\frac{1}{3}},$$

where λ is hardness of the silo.
- The hardness of the Soviet silo, λ, was denoted as the lethal radius of a single 1-megaton warhead. Thus, if a 1-megaton warhead had a lethal radius of 1,000 feet, an 8-megaton warhead would have an LR of 2,000 feet, since

$$\sqrt[3]{8} = 2.$$

Given these conditions, the P_s of a Soviet silo from an attack by a single warhead was

$$P_s = 0.5^{\frac{\lambda^2 y^{2/3}}{C^2}}.$$

Then, the P_s from n RVs against one silo would be

$$P_s = 0.5^{\frac{n\lambda^2 y^{2/3}}{C^2}}.$$

If we demand a DE of 0.9, it means that the exponent for 0.5 in the equations must have the value of 3.32 (since $0.5^{3.32} = 0.10$).

Now we can gain some insight as to the relative contributions to the total HTK of each of the measures proposed:

- If you increase the yield of a warhead from 250 kt to 400 kt (one of BMO's proposals), you will increase the value of the exponent by a factor of 1.37, the calculation being

$$\left(\frac{400}{250} \right)^{\frac{2}{3}} = 1.37.$$

- If you decrease the CEP by a factor of 1.2, you will increase the value of the exponent by a factor of 1.44, that is, $(1.2)^2 = 1.44$.
- You can gain the same increase in the value of the exponent (namely, a factor of 1.44) by deploying 1.44 times as many deliverable RVs.

The first two measures (increasing the yield and decreasing the CEP) resulted in a combined increase of HTK by a factor of 1.97. In their presentation to the Air Staff, the BMO staff packaged the two measures together and stated that their proposal would "leverage the force by almost double." But they were silent about the relative contribution of each measure. The total bill for their proposed program was, even by BMO's own estimates, quite large, and I believed that their estimates were grossly understated.

General Ryan asked me to evaluate BMO's proposal. In my briefing to him I confirmed that increasing the yield of the Minuteman warhead from 250 to 400 kt and reducing the CEP by a factor of 1.2 would, indeed, double the HTK of the Minuteman force. But I went on to point out that the majority of this increase in HTK stemmed from improving the guidance system, while the great majority of the cost of BMO's proposal came from developing and deploying a new RV. This was so because many launches of the missile with the new warhead would be needed to test the RV's ballistic properties and because large amounts of enriched uranium would have to be procured. According to my figures, the ratio of costs (of the new RV versus the new guidance system) was six to one. Based on my analysis, the Chief of Staff and the Secretary of the Air Force approved a program to develop the new guidance system for the Minuteman, but they did not approve a program for the new RV.

The above demonstrates that simple constructs can provide insight and reliably inform decisions about whether or not to proceed to implement some concept being proposed.

This episode was but one of a continuing battle between some members of the Air Staff and the ballistic missile program office. BMO continually sought funding to proceed with the development of a new RV for Minuteman, and I prevailed on the issue throughout my tenure in AFSA. However, subsequent to my departure from AFSA, a new Chief of Staff did grant approval to proceed with a new RV. The program was hardly a success—a large cost overrun occurred, and the resulting RV failed to meet its expected performance specification with respect to its yield.

The Trade-Off with "Soft" Area Targets

Now examine the trade-off between yield and numbers for the case of attacking "soft" area targets. Such targets include industrial infrastructure and unhardened military targets. In this case, the trade-off between numbers and yield is not so obvious. If the area occupied by the target is very large compared to the lethal area of one weapon, then the trade-off is the same: $ny^{2/3}$, where n is the number of RVs and y is the yield of each. But that is seldom the case; industrial facilities are generally built not in large, circular clusters but rather more on a line (e.g., along a river or railroad within a valley). If the facilities lie in a line whose width is less than the diameter of the lethal area of the weapon, then we can, by math, announce the trade-off: $ny^{1/3}$. The one-third term in the exponent results from the fact that some of the weapon's effects are expended outside of the target area. We have now bounded the problem: The exponent of y is somewhere between 0.33 and 0.67.

One might be tempted to take the arithmetic mean between these two values—0.5—but there would be a hue and cry about mathematical inelegance. So I devised a more complicated method for arriving at the answer. Specifically, I devised a chart with "Soviet value destroyed" on the ordinate and "number of weapons" on the abscissa and, for a family of curves, yields of 100, 200, 500, 1,000, and 2,000 kt. It took some effort to construct the lines, but finally we had the chart. Obviously, all lines (one line for each yield) started at zero and were concave downward—reflecting the fact that, as you went to more and more weapons on the abscissa, you were attacking targets of less and less value and thus for diminishing returns. Obviously, the 1-megaton lines rose more rapidly than the lines for lesser yields.

Now the trick: Draw a horizontal line from some place midway up the ordinate. This is a line of constant value destroyed. For example, we note that, for the line of 200 kt, it took 800 weapons to achieve this level of damage, and that, for the 1-megaton line, it took only 360 weapons. Now find the value of z so that $0.200^z \times 800$ is equal to $1^z \times 360$. The answer is $z = 0.5$. That is, when the exponent is 0.5, the square root of $0.200 \times 800 = 358$—close enough. Obviously, I took some other numbers and did not get the same value for z for all pairs. But the average value for the exponent was about 0.5—maybe a little less.

I now have stored in my mind that the exponent to use in determining the trade between numbers of weapons (n) and the yield of each (y) is $Y^{0.5}$ for the case of soft area targets. As we saw earlier, this relationship comes in handy in gaining insight into problems such as determining the number of RVs on the front end of the Minuteman III missile (see pp. 144–146).

Calculus of the Attrition of Agents in a Battle

Late in fall 2002, several RAND analysts participated in the Air Force's biennial war game, Global Engagement. Following the game, one of them, David Ochmanek, who had played on the Blue team during the game, approached me.

He said that, on the third day of the game, the players had been working through the scenario. Late in the afternoon, both the Blue and the Red teams were instructed to commit to a course of action. At this point in the game, it was clear that there would be a clash of forces. The Blue team had defined the types and numbers of forces it wanted to employ, as well as where and under what conditions they were to engage. The team assumed that Red had done the same for its forces. To proceed with the scenario, both sides needed to know the outcome of the battle that would ensue. That is, they needed to know the "fraction surviving" for each type of force or agent involved in the battle after a stated period of time, say, 24 hours.

The control team for the game gave this problem to a team of analysts whose job it was to assess such interactions so that control could adjudicate each major move by the teams. The analysts, armed with databases and computer models, worked feverishly until after midnight and provided the Blue team with some answers the following morning. After some critical review of their product, the team came to the conclusion that the results from the analysts could not stand the light of day and could not be used. The control team essentially had to make up the outcome of the move.

Ochmanek then issued a challenge by asking whether there wasn't a way to provide assessments of the outcomes of a stated battle that were quick (require less than an hour to generate); transparent; coherent; and, at the same time, rigorous. I told Dave I would attempt to define such a calculus. After some false starts and less-than-useful digressions, I was well on my way to defining such an approach when fate intervened. I became a resident of the Walter Reed Army Hospital for some six months. After my return home, I thought about the problem episodically. I also talked with Dr. Leon Goodson and Dr. Scott Meyer of STR—both of whom had worked for me when I ran AFSA. They provided some helpful insights and suggestions. Dr. Bob Sheldon, also formerly of AFSA, was also of great help in defining the calculus. After a time, I took the opportunity to discuss this whole affair with Dr. Jacqueline Henningsen, the Air Staff's Director for Studies and Analyses, Assessments and Lessons Learned. She was interested in pursuing the matter and assigned Maj Michael Kram to fashion a briefing describing the calculus I had proposed. What follows is a summary of the approach I defined.

Basic Principles

Again, our objective was to estimate the outcome of a battle, which generally involves multiple types of Blue and Red agents, and to do so in a way that is rigorous, transparent, and quick. The approach I developed meets these requirements. It is based on a summing up of the outcomes of the types of engagements (agent-on-agent interactions) the battle is likely to comprise. The calculus defines the fraction surviving of each type of agent over time. The fraction surviving Red (FSR) is the fraction of Red agents of a particular type that remain after a time step (Δt) has elapsed[4]:

$$FSR = \exp\left(\frac{-\gamma B \Delta t}{R}\right),$$

where

B = the number of Blue agents at the beginning of the time step

R = the number of Red agents at the beginning of the time step

Δt = a small time step (e.g., one-tenth of an hour)

γ = the kill potential per hour of a particular type of Blue agent against a particular type of Red agent.

FSR can be simplified as

$$FSR = \exp\left(-\Omega_R \Delta t\right),$$

where Ω_R represents the weighted sum of all the kill potentials arrayed against the Red agents in question. In this way, the analyst can calculate the result for a large number of engagements by using this expression for the overall "stress" placed on a particular type of Red agent from all the Blue agents that are engaging that type. Calculating the fraction surviving Blue (FSB) is done in an analogous way. The γ term

---

[4] It is important that the drawdown calculations be accomplished in small "bites" (that is, that Δt be of short duration) so that the average population of Red and Blue agents over the course of each time step be roughly equal to the population at the beginning of the time step.

in the equation above is, in this case, replaced by λ, which represents the kill potential per hour of a particular type of Red agent against the relevant type of Blue agent.

Obviously, the key to the validity of this approach is to have credible values for γ and λ. This is not simple but it is tractable. Values for these variables can be developed in any of five ways:

- using the judgment of operators and analysts informed by data from past engagements in actual conflicts
- performing analyses using inputs from physics, engineering, and mathematics
- conducting analyses using the outputs of such high-fidelity engagement-level simulators as TAC BRAWLER
- conducting field trials and analyzing their results
- some combination of the above approaches.

To illustrate this, we consider how an engagement-level simulator might be harnessed to derive kill potentials for two types of aircraft, one Red and one Blue. The analyst would be asked to set up a run of the simulator in which a modest number of Red and Blue platforms engage in "combat," say, four Su-27 and four F-22 fighters. The simulator is run until the *FSR* or the *FSB* reaches a predefined level. In this case, we will halt the simulation when the *FSR* equals 0.5, meaning that half the Red aircraft (i.e., two of them) have been "killed." At this point, we observe two other types of data from the simulator: the time (in "simulation time") at which this threshold is reached and the FSB—the fraction of Blue aircraft surviving.

Armed with these data, we can calculate γ and λ for this pair of agents:[5]

$$\gamma = \left(-\ln FSR\right)\left[\frac{R_0\left(1 + FSR\right)}{B_0\left(1 + FSB\right)\Delta t}\right]$$

---

[5] Note that the terms (1 + *FSR*) and (1 + *FSB*) are inserted into the equations to capture the fact that the average population of *B* and *R* during the time step Δt is different from R_0 and B_0.

and

$$\lambda = \left(-\ln FSB\right)\left[\frac{B_0\left(1 + FSB\right)}{R_0\left(1 + FSR\right)\Delta t}\right].$$

The analyst then repeats the experiment many times (assuming a stochastic model) and calculates the median values of γ and λ accordingly. When this is done to the satisfaction of all concerned, the resulting values of γ and λ are stored in a catalog. The process is then repeated for each pairing of types of agents that could occur in a conflict. So, if Red operates Su-27s, Su-30s, F-9s, and F-10s and if Blue operates F15Cs, F-15Es, F-16 Block 50s, F-22s, and F/A-18E/Fs, we will need to run the simulator for all 20 possible pairings.

Scaling Up

Before we can assess a "real" potential conflict, however, we must go further. A war between the United States and China, for example, could be expected to result in large-scale battles. It might not be uncommon for, perhaps, 40 Blue agents of one type to engage 40 Red agents of another type. Yet our simulator can handle only small engagements, perhaps as large as four on four. Fortunately, the calculus can be scaled up to handle the larger battle. We do this by using the values for γ and λ that were derived as outlined above and plugging them into the equations for FSR and FSB, as shown below. Assume, for the purposes of illustration, that the median values from our simulator runs are $\gamma = 0.2669$ and $\lambda = 0.0347$. Assume also that B_0 and R_0 (the number of Blue and Red agents at the outset of the battle, when $t = 0$) are both equal to 40:

$$FSR = \exp\left(\frac{-0.2669 \times 40 \times 0.1}{40}\right) = 0.9737$$

and

$$FSB = \exp\left(\frac{-0.0347 \times 40 \times 0.1}{40}\right) = 0.9965.$$

Note that here Δt = 0.1 hours, or 6 minutes. We know, however, from our previous work with the engagement-level simulator, that it typically took considerably more time (in our example, 2 hours) to reach FSR = 0.5. (Experience also suggests that a 40-on-40 engagement would last much longer than 6 minutes.) We therefore iterate these equations in successive six-minute time steps until we reach the two-hour point. The results of these calculations are shown in Table 6.1. As before, the FSR is approximately one-half (in this case, 0.4902), meaning that approximately 20 Red aircraft are assessed to have been destroyed in this air battle. The FSB after two hours is 0.9477, meaning that approximately two Blue aircraft have been destroyed $(1 - 0.9477) \times 40$ = 2.092. This, then, is the estimated drawdown of Red and Blue agents after two hours of battle: 20 Red survivors and 38 Blue.

Heterogeneous Engagements

The final step is to apply this approach to the assessment of "heterogeneous" battles in which platforms engage dissimilar types of platforms. For example, we know that SAMs can engage aircraft and vice versa. Likewise, multirole and ground-attack aircraft can engage enemy tanks, although tanks do not typically engage aircraft. How can these sorts of engagements be handled?

We begin by introducing a standard format for displaying the relevant data. Figure 6.4 shows the information the analyst needs to assess the outcome of a battle in which the Red commander decides to apportion three types of forces in the following way:

- 90 MiG-29s will be apportioned to intercept an attack of F-15Es expected to number 200.
- 50 SA-99 SAMs will be directed to engage both F-15Es and F-16 aircraft that come into range. Given the proportion of both types of aircraft in the Blue order of battle and other factors, it is anticipated that 30 of the SA-99s will engage F-15Es and 20 of them will engage F-16s.
- 500 T-84 tanks are directed to engage the enemy's armored force—300 M-1 tanks.

Table 6.1
Tabular Data from 40-on-40 Example

Time Step	FSR	FSB
0.1	0.9737	0.9965
0.2	0.9474	0.9932
0.3	0.9213	0.9899
0.4	0.8953	0.9867
0.5	0.8693	0.9836
0.6	0.8434	0.9806
0.7	0.8177	0.9777
0.8	0.7920	0.9748
0.9	0.7664	0.9721
1.0	0.7409	0.9694
1.1	0.7155	0.9669
1.2	0.6901	0.9644
1.3	0.6649	0.9620
1.4	0.6397	0.9597
1.5	0.6146	0.9575
1.6	0.5895	0.9554
1.7	0.5646	0.9533
1.8	0.5397	0.9514
1.9	0.5149	0.9495
2.0	0.4902	0.9477

Note that, in Figure 6.4, Red forces are arrayed across the top and Blue forces are along the left-hand side.

The number in the center of each box above represents λ—the kill potential per hour that the agent at the top of the column can inflict on the agent named at the far left-hand side of the row. So, in the upper left-hand box, we show that a MiG-29 is judged to be able to kill, on average, 0.15 F-15Es in 1 hour of combat. The number in the bottom right-hand corner of each box represents the stress factor,

ω_B, imposed by the apportioned Red agents of a given type against the total number of Blue agents they are apportioned against.

$$\omega_B = \frac{\lambda R}{B}.$$

So, for example, looking again at the upper left-hand box, the stress factor imposed by 90 MiG-29s on 200 F-15Es is

$$\frac{0.15 \times 90}{200} = 0.0675.$$

The total stress factor (or kill potential per hour) each type of Blue agent faces (Ω_B) is listed on the right-hand side of each row and is derived simply by summing the stress factors of each of the Red

Figure 6.4
Red Operational Plan

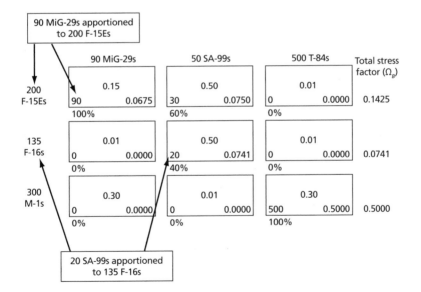

NOTE: Operational factors are notional.
RAND *OP223-6.4*

agents in the row. For example, Blue's F-15E force will face a total stress factor,

$$\Omega_B = 0.0675 + 0.0750 + 0.0000$$
$$= 0.1425.$$

The Blue commander prepares his OPLAN in a similar manner, as shown in Figure 6.5:

- He sends 180 F-15Es to engage the enemy's force of MiG-29s. The remaining 20 F-15Es are to engage the enemy's SA-99 SAMs.
- The F-16s are to try to avoid Red's MiG-29s. Instead, 50 of the F-16 sorties are to engage the SA-99s, and the remaining 85 are to engage Red's tanks.
- All 300 of Blue's M-1 tanks are to engage Red's tanks.

Figure 6.5
Blue Operational Plan

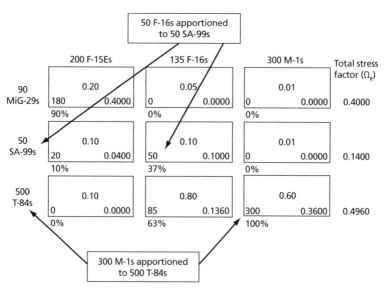

NOTE: Operational factors are notional.
RAND OP223-6.5

As before, the total stress factor each Red component faces (Ω_R) is shown at the far right.

Using the equations shown above, we can readily calculate the FSRs and FSBs for this battle. Again, we will run the "model" in six-minute time steps for a total of two hours of "scenario time." The result is the drawdown of Blue and Red forces shown in Figures 6.6 and 6.7.

Future Applications

The approach outlined here could be useful to analysts of defense problems in many ways. It combines several features that make it attractive as a tool for analysis, namely, the high fidelity of engagement level simulations, the simplicity and transparency of a fast-running, spreadsheet-based campaign-level model, and a way of arranging the inputs (numbers and types of agents and their apportionment among mission areas) to reflect the way operators think about conducting battles. This approach aggregates assessments of engagements into an assessment of a battle. By "resetting" the apportionments in 12- or

Figure 6.6
The Drawdown of Blue Forces

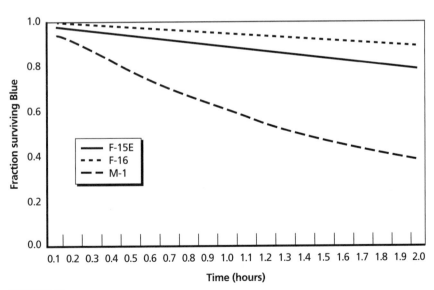

Figure 6.7
The Drawdown of Red Forces

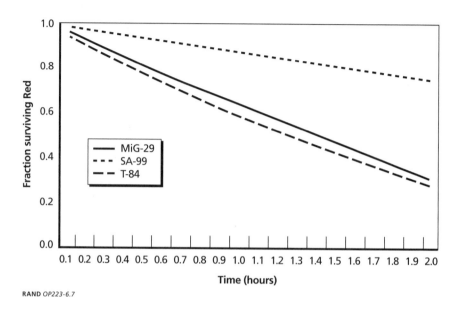

24-hour time steps, analysts can aggregate their assessments of a series of battles into the assessment of a multiday campaign.

Accordingly, in presentations to members of the defense analytic community, I have recommended adoption of this approach for use not only in adjudicating the outcome of moves in war games but also in addressing a wide range of force-planning issues. Force planners need to be able to understand how changes in force size and modernization will affect the operational capabilities of the armed forces and their ability to accomplish future missions. This tool can shed considerable light on such questions. Because this approach is "user friendly" to operators, Joint Force Commanders and others charged with formulating strategies for the conduct of specific campaigns could also use it as a means of evaluating and comparing alternative courses of action.

As noted above, the key to making this tool work is generating credible and widely accepted values for the kill potential of every major agent type likely to be employed in theater conflict. These must be developed for every type of target that each agent might engage. The result would be a "catalog" of kill potentials for all potential pairings of shooters and targets. This is not new: Several campaign models that have been in use over the past decades have similar embedded killer-victim scoreboards. When these values have been based on sound engagement-level field tests or simulations, they may be readily adapted to this approach. When such values are not available, new sets of simulator runs will be required. Ideally, such an effort would be organized and managed centrally by a joint entity, such as OSD PA&E or the Joint Staff's Force Structure Resources and Assessment, so as to spread the burden and help ensure that the resulting values would be accepted and used across the services and commands.

Once the catalog was constructed, the analysts charged with assessing the outcome of battles in support of large war games would be able to do so quickly and with high levels of fidelity and credibility. They would know in advance what types of weapon systems all sides would employ in the game. Armed with this knowledge, they could set up their spreadsheets in advance and then select the appropriate

---

Box 6.1
An Opportunity Lost

As a sidebar to this episode, I had used this approach in the 1960s in the calculus for determining the fraction of bombers that would be expected to penetrate an enemy defense (see "Penetrating Soviet Air Defenses: The Argument for Decoys," pp. 149–153). That is, we were able to determine the kill potential of Red SAMs and aircraft interceptors against Blue bombers using the output of a simulator called the Advanced Penetrator Model. These values along with the "exponential" equation, enabled us to determine the fraction surviving Blue (FSB) for the bombers.

To this day, I regret that I did not pursue this approach further and apply it to the battle on the ground. I could have done this while head of AFSA and, more to the point, when I was the head of WSEG. While at WSEG I did let three contracts intended to define what we called a "campaign model." (This effort is chronicled in "Fostering Campaign Models," pp. 223–226.) In hindsight, I could have done much better and was, as head of WSEG, in a position to enforce the use of such an approach—an opportunity lost.

apportionments based on instructions they receive from each team. The result would be a win-win situation: Game sponsors and players would get timely, credible assessments of the outcome of game moves, and the analysts, for once, could enjoy a good night's sleep.

CHAPTER SEVEN

Summing Up: Kent's Maxims

We conclude with a distillation of the major lessons that General Kent has drawn from his experience of more than half a century of service to the nation. These lessons, or maxims, emerge from the stories related in the preceding chapters and are relevant for everyone engaged in the defense policy process.

Creating Effective Analyses

Think Before You Calculate

Devising the basic analytic construct to apply to a problem is far more important than crunching numbers. Scope the problem carefully, and address the key assumptions, instead of rushing to gather data and doing calculations.

In other words, just sit back and think. Often, doing this can reveal a basis for calculations that are quite straightforward and that yield new insights into the most important aspects of the problem.

Minimize Reliance on Computers

Used appropriately, computers can be invaluable tools, but they can also hide a multitude of errors. It is often best to do your calculations and plots by hand, particularly at first, until you are sure that you fully understand the interactions you are examining and have wrung all the "bugs" out of your methodology.

Seek Help from Outside Experts

When confronted with a "new" problem, cast a wide net to determine whether someone else has already solved it. More broadly, consult early and often with the most qualified people you can find to help you understand the problem.

Do Not Treat the Adversary as Static

In military affairs, as in most fields of human endeavor, opponents react to each other's moves. Although this seems obvious, it is surprisingly common for advocates of certain policies or programs to assume that the adversary will not react to our initiatives.

Eschew "Recommendations"

The purpose of analysis is to provide illumination and visibility—to expose some problem in terms that are as straightforward as possible. If the analyst can illuminate the problem for decisionmakers, they can decide what course of action to pursue without getting formal "recommendations" from the analyst.

Recruit People Who Can Think

Analysts should be recruited because they have the talent to dissect problems—to collapse seemingly complicated phenomena into much simpler constructs.

These individuals are to be graded more on impeccable logic than on correct arithmetic. They are to be graded as well on how elegantly and simply they are able to "model" (in the broad sense of the word) some problem.

Invest in People

If you are running an organization, invest plenty of time and effort in recruiting the best people you can get and in developing them.

The best education for an analyst is in the school of doing. Don't be afraid to give people challenging problems to work on.

Use No-Holds-Barred "Murder Boards" to Improve Your Products

A briefer should face the toughest scrutiny from the internal reviews of his or her work, not from outsiders.

Encourage your people to disregard hierarchy. The lowest-ranking person in the organization should feel free to challenge the boss on any point.

Making (Good) Things Happen

Convene Conceivers Action Groups to Promote Innovation in Operational Capabilities

The best way to solve a complex operational problem is to frame it clearly and convene an interdisciplinary group to tackle it

Clearly stating the challenge is half the battle. Concept development must be an exercise in solving a problem, not in "studying" a problem or some set of technologies. The problem to be solved should generally be framed as a discrete operational task or objective (a set of related tasks).

Give the challenge to a group made up of individuals with multiple competencies, including relevant technologies, military operations, enemy capabilities, system development, engineering, and analysis.

Beware of the "Hobby Shop" Mentality

Too many people at government laboratories are more interested in maturing technologies than in putting rubber on the ramp. They must be reminded from time to time that it is, ultimately, fielded capabilities that matter to our commanders and forces.

Draw a Bright Line Between Demonstrating a Technology and Developing a System

The company that demonstrates a technology is not always the one best suited to developing and producing a system incorporating that technology. Generally, the contract to do the latter must be competed, with the source-selection authority at a fairly high level. Attempting to short-circuit this process will lead, more often than not, to challenges to the program and needless delays.

Be an Advocate

Just having a good idea or being right about something does not ensure success. You often have to put as much effort into communicating and proselytizing your ideas as you do into developing them:

- Have a well-thought-out plan of action.
- Effective advocacy begins with showing how your concept provides an important new capability.
- It pays to have friends in high places. It is great to be right, but merely being right is no guarantee against getting into hot water. At such times, having the support of a high-ranking official is invaluable.
- Anticipate decisionmakers' key questions and address them before they are even asked.
- Recognize that important decisions often revolve around personalities more than formal documentation.
- The best way to generate high-level support for new concepts is to sell the user—in the case of operational concepts, the combatant commander—on them. Go to the top, and show the four-star how your concept can solve a problem of great importance to him.

Doing the Right Thing

To Thine Own Self Be True

Do not accept direction simply because of the authority of its source. If you are confident in the integrity of your analysis (and you should be), be prepared to go broke on your own strengths and weaknesses, not the dictates or prejudices of those who may be (in an organizational sense) your superiors.

Beware of Statements of "Operational Requirements"

In developing new systems, adhering mindlessly to statements of so-called operational requirements can be fatal. Such statements are often defined rather arbitrarily by people who have little appreciation either for what is feasible or for the trade-offs involved in creating a workable system.

Defining a new system should be a cooperative endeavor between engineers and operators: The engineers define the limits of technology; the operators define the best balance of characteristics within those limits. Since the "best balance" of characteristics must necessarily be determined with associated costs in mind, someone with an idea of the intrinsic costs associated with the key characteristics should be involved as well.

Accept Risks

This applies personally as well as institutionally. Highly risk-averse people rarely accomplish much. Institutions willing and able to take calculated risks can make big strides in capabilities.

Winning Bureaucratic Battles

Understand the Dynamics of the Real Decisionmaking Process

Bureaucratic wiring diagrams seldom reflect the reality of how decisions are made. Focus on people and decisions by people, and figure out how to inform the decisionmaker so he or she will make the best choice.

Go to the Top

Whenever possible, avoid wasting time arguing with people who do not have the authority to act or by filling squares as dictated by pointless regulations.

Seize the Conceptual High Ground

When you set out to change a policy, seek to define the issue in such a way that arguments against your position are simply untenable.

Anticipate the Need for Analysis

Whenever possible, you should strive to get your analysis and its implications injected into the policy debate early, before minds are made up and positions are set. Often, this means getting an analysis started long before the issue "heats up."

Recognize that a Good Offense Is Usually Better Than a Good Defense

It is more fruitful to attack the critique of your analysis by others than to try to prove that your analysis is without error:

- If you undertake to challenge or discredit a report, focus and dwell on the one or two points on which its authors are obviously wrong, and on which you can prove they are wrong.
- Identify the arguments that you have to win.
- Isolate and dwell on one key issue, sentence, number, or word.

Encourage Errors by Your Adversary

Unless you are confident that you can decisively shape the basic results of a "joint" study, it's a fool's errand to seek incremental changes to it. Better to let the people in charge do their thing and hope that their work is fatally flawed, while you, at the same time, independently conduct the study that ought to be done.

And Finally: Do Your Homework

There is simply no substitute for being the smartest person in the room about the issue at hand.

Chronology

June 25, 1915	**Glenn Altran Kent** born Red Cloud, Nebraska
1918	Family moves to Manzanola, Colorado
1932	Graduates as high school valedictorian
1936	Graduates with major in mathematics Western State College Gunnison, Colorado
1936 through 1941	Teaches high school math and chemistry Hotchkiss, Colorado
June 1941	Joins the Army Air Corps
July 1, 1941	Aviation cadet California Institute of Technology Pasadena, California
1942	Receives master's degree in meteorology California Institute of Technology Pasadena, California
February 13, 1942	Commissioned second lieutenant, Army Air Corps Attended boot camp March Field, California

March 1942	On detached service to Eastern Airlines, Hopeville, Georgia
1942	Promoted to first lieutenant
July 1942	Assigned to weather station Goose Bay, Labrador
Spring 1943	Promoted to captain
July 15, 1943	Station Weather Officer and Chief of Weather Station BW-1 Narsasuak, Greenland
1944	Promoted to major
1945	Chief of Weather Station, Grenier Field, New Hampshire
January 1946	Discharged from Army Air Corps
April through December 1946	Employed by Bureau of Reclamation Denver, Colorado
December 17, 1946	Called back to serve in the Army Air Corps
1947	On station Goose Bay, Labrador
September 1947	On station Westover Field, Massachusetts
October 1947	Studied math, physics, and radiological engineering Naval Postgraduate School Annapolis, Maryland
June 1948	Studied radiological engineering University of California Berkeley, California

July 1950	Armament Division Directorate of Research and Development Headquarters, U.S. Air Force
October 19, 1950	Promoted to lieutenant colonel
1953	Married Phyllis Horton of Richlands, Virginia
1953	Air Force Special Weapons Center Kirtland Air Force Base, New Mexico Last position was deputy to the director of research
1955	Promoted to colonel
1956 through 1957	Student Air War College Maxwell Air Force Base, Alabama
1957	Chief Weapons Plans Division Directorate of Plans Headquarters, U.S. Air Force
1961	Fellow Center for International Affairs, Harvard University

1962	Military Assistant to the Deputy Director
	    (Strategic and Defensive Systems)
	    Defense Research and Engineering
	Office of the Secretary of Defense

1963	Promoted to brigadier general
July 1965	Assistant for Concept Development to the Deputy Chief of Staff for Development, Headquarters, U.S. Air Force

September 1966	Chief of Development Planning Headquarters, Air Force Systems Command Andrews AFB, Maryland
1966	Promoted to major general
August 1968	Assistant Chief of Staff Air Force Studies and Analysis Headquarters, U.S. Air Force Reported directly to the Air Force Chief of Staff
1972	Promoted to lieutenant general

February 1972	Director
	Weapon System Evaluation Group
	Washington, D.C.
	Reported to both the Chairman of the
	    Joint Chiefs of Staff and the Director
	    of Defense Research and Engineering,
	    Office of the Secretary of Defense

September 1974	Retired from active duty in the U.S. Air Force
1974 through 1982	Consultant to various defense contractors
1982 to the present	Senior research fellow The RAND Corporation

Awards

Defense Distinguished Service Medal

U.S. Air Force Distinguished Service Medal with Oak Leaf Cluster

Legion of Merit with Oak Leaf Cluster

National Defense Service Medal

European–African–Middle Eastern Campaign Medal

Commendation Medal with Oak Leaf Cluster

Department of Air Force Decoration for Exceptional Civilian Service (twice)

U.S. Air Force Analysis Community Lifetime Achievement Award

Glenn A. Kent Leadership Award (created)

Vance R. Wanner Memorial Award

Jacinto Steinhardt Memorial Award

Bibliography

Birkler, John, C. Richard Neu, and Glenn A. Kent, *Gaining New Military Capability: An Experiment in Concept Development*, Santa Monica, Calif.: RAND Corporation, MR-912-OSD, 1998. As of March 17, 2008:
http://www.rand.org/pubs/monograph_reports/MR912/

Coram, Robert, *Boyd: The Fighter Pilot Who Changed the Art of War*, New York: Little, Brown, and Co., 2002.

Department of Defense Directive 5000.1, *The Defense Acquisition System*, May 12, 2003.

Directorate of Defense Research and Engineering, *A Summary Study of Strategic Offensive and Defensive Forces of the U.S. and USSR*, Washington, D.C., September 8, 1964.

Finn, Michael V., and Glenn A. Kent, "Simple Analytic Solutions to Complex Military Problems," Santa Monica, Calif.: RAND Corporation, N-2211-AF, 1985. As of March 17, 2008:
http://www.rand.org/pubs/notes/N2211/

Graham, Daniel O., and Gregory A. Fossedal, "A Defense That Defends," *Wall Street Journal*, April 8, 1983.

Headquarters, Strategic Air Command, "History and Research Division, History of the Joint Strategic Planning Staff: Background and Preparation of SIOP-62," n.d. As of February 13, 2008:
http://www.gwu.edu/~nsarchiv/NSAEBB/NSAEBB130/SIOP-28.pdf

Iklé, Fred Charles, "Nuclear Strategy: Can There Be a Happy Ending?" *Foreign Affairs*, Vol. 63, No. 4, Spring 1985, pp. 810–826.

Joint Secretariat, "Review of the Initial NSTL and SIOP," note to the Joint Chiefs of Staff on JCS 2056/194, December 9, 1960. As of February 13, 2008:
http://www.gwu.edu/~nsarchiv/NSAEBB/NSAEBB130/SIOP-18.pdf

Kent, Glenn A., "On the Interaction of Opposing Forces Under Possible Arms Control Agreements," Cambridge, Mass.: Harvard University Center for International Affairs, Occasional Paper No. 5, March 1963.

————, "Concepts of Operations: A More Coherent Framework for Defense Planning," Santa Monica, Calif.: RAND Corporation, N-2026-AF, 1983. As of March 17, 2008:
http://www.rand.org/pubs/notes/N2026/

————, "Decision-Making," *Air University Review*, Vol. XXII, No. 4, May–June 1971, pp. 62–65.

————, "On Analysis," *Air University Review*, Vol. XVIII, No. 4, May–June 1967, pp. 50–54.

————, "A Suggested Policy Framework for Strategic Defenses," Santa Monica, Calif.: RAND Corporation, N-2432-FF/RC, 1986. As of March 17, 2008:
http://www.rand.org/pubs/notes/N2432/

Kent, Glenn A., and Randall J. DeValk, *Strategic Defenses and the Transition to Assured Survival*, Santa Monica, Calif.: RAND Corporation, R-3369-AF, 1986. As of March 17, 2008:
http://www.rand.org/pubs/reports/R3369/

Kent, Glenn A., Randall J. DeValk, and David E. Thaler, "A Calculus of First-Strike Stability: A Criterion for Evaluating Strategic Forces," Santa Monica, Calif.: RAND Corporation, N-2526-AF, 1988. As of March 17, 2008:
http://www.rand.org/pubs/notes/N2526/

Kent, Glenn A., Randall J. DeValk, and Edward L. Warner III, *A New Approach to Arms Control*, Santa Monica, Calif.: RAND Corporation, R-3140/FF/RC, 1984. As of March 17, 2008:
http://www.rand.org/pubs/reports/R3140/

Kent, Glenn A., and David A. Ochmanek, *A Framework for Modernization Within the United States Air Force*, Santa Monica, Calif.: RAND Corporation, MR-1706-AF, 2003. As of March 17, 2008:
http://www.rand.org/pubs/monograph_reports/MR1706/

Kent, Glenn A., and David E. Thaler, *First-Strike Stability: A Methodology for Evaluating Strategic Forces*, Santa Monica, Calif.: RAND Corporation, R-3765-AF, 1989. As of March 17, 2008:
http://www.rand.org/pubs/reports/R3765/

————, *First-Strike Stability and Strategic Defenses: Part II of a Methodology for Evaluating Strategic Forces*, Santa Monica, Calif.: RAND Corporation, R-3918-AF, 1990. As of March 17, 2008:
http://www.rand.org/pubs/reports/R3918/

Reagan, Ronald, "Address to the Nation on Defense and National Security," March 23, 1983. As of October 26, 2006:
http://www.reagan.utexas.edu/archives/speeches/1983/32383d.htm

Schelling, Thomas C., *The Strategy of Conflict*, Cambridge, Mass.: Harvard University Press, 1960.

————, "What Went Wrong with Arms Control?" *Foreign Affairs*, Vol. 64, No. 2, Winter 1985–1986, pp. 219–233.

Talbott, Strobe, *Deadly Gambits: The Reagan Administration and the Stalemate in Nuclear Arms Control*, New York: Knopf, 1984.

Thaler, David E., *Strategies to Tasks: A Framework for Linking Means and Ends*, Santa Monica, Calif.: RAND Corporation, MR-300-AF, 1993. As of March 17, 2008:
http://www.rand.org/pubs/monograph_reports/MR300/

Twining, Nathan, Chairman of the Joint Chiefs of Staff, to the Secretary of Defense, "Target Coordination and Associated Problems," August 17, 1959. As of February 13, 2008:
http://www.gwu.edu/~nsarchiv/NSAEBB/NSAEBB130/SIOP-2.pdf

U.S. Code, Title 10, Armed Forces, Chapter 803, Department of the Air Force, January 19, 2004.

Warner, Edward L., III, and Glenn A. Kent, "A Framework for Planning the Employment of Air Power in Theater War," Santa Monica, Calif.: RAND Corporation, N-2038, 1984. As of March 17, 2008:
http://www.rand.org/pubs/notes/N2038/

White, Thomas, Air Force Chief of Staff, memorandum to Secretary of Defense Thomas Gates, with attachment on Strategic Targeting Authority, June 10, 1960. As of February 13, 2008:
http://www.gwu.edu/~nsarchiv/NSAEBB/NSAEBB130/SIOP-6.pdf

Wilkening, Dean, and Kenneth Watman, *Strategic Defenses and First-Strike Stability*, Santa Monica, Calif.: RAND Corporation, R-3412-AF, 1986. As of March 17, 2008:
http://www.rand.org/pubs/reports/R3412/